CW00696064

The Genus Dionysia

1 *Dionysia diapensiifolia* in cultivation; pin-eyed plant showing exserted styles. Photo. J. G. Elliott.

2 *Dionysia involucrata* growing on the limestone cliffs of the Varzob Gorge, S. USSR.
Photo. R. B. Burbidge.

The Genus Dionysia

CHRISTOPHER GREY-WILSON

Royal Botanic Gardens, Kew

ALPINE GARDEN SOCIETY

First Published 1989

© C. Grey-Wilson

Published by Alpine Garden Society
Lye End Link
St John's
Woking
Surrey GU21 1SW

Editor: Richard Bird
Assistant Editor: Robert Rolfe
Designer: John Fitzmaurice

ISBN 0 900048 51 4

Typeset by Tradespools Limited, Frome, Somerset

Printed by Jolly and Barber, Rugby

Colour by Waterden Reproduction, London

ACKNOWLEDGEMENTS

I am very much indebted to many people for help and advice in producing this new *Dionysia* monograph. Their knowledge and expertise has been an inspiration to me, and I trust that they approve of the way that I have tried to amalgamate the various sources of information. I must include here the following: Jim Archibald, Brian Burrow, Jack Elliott, Alf Evans, Harold Esselmont, Tony Hall, Tom Hewer, Z. Jamzad (Tehran), Brian Mathew, Bob Mitchell, Chris Norton, Mary Randall, Geoff Rollinson, Henry Taylor, Stan Taylor, Eric Watson and Henrik Zetterlund.

During the preparation of my initial book on Dionysias for the Alpine Garden Society (1970) I was given a good deal of help form several well-known people who are now sadly dead – E.K. Balls, Peter Edwards, Paul Furse and Per Wendelbo. This present work would have been impossible without their generous advice, which formed the 'core' of the earlier publication.

Contents

List of Colour Plates

This monograph is dedicated to Professor Per Wendelbo (formerly of Göteborg University) who did so much of the detailed research on the genus, and who, over a number of years, stimulated interest in the species.

3 *Primula verticillata*. Photo.
C. Grey-Wilson.

4 *Primula floribunda*. Photo. C. Grey-Wilson.

5 *Dionysia mira* in cultivation at RBG Kew. Photo. C. Grey-Wilson.

6 *Dionysia teucrioides*. Photo. J. G. Elliott.

7 *Dionysia balsamea* growing on limestone cliffs at Djam, Afghanistan. Photo. C. Grey-Wilson.

8 *Dionysia balsamea*. Photo. S. Taylor.

9 *Dionysia paradoxa* growing on conglomerate cliffs, Sarobi Gorge, Afghanistan. Photo. C. Grey-Wilson.

10 *Dionysia aretioides* growing on limestone cliffs of the Chalus Gorge, N. Iran.
Photo. B. Mathew.

11 *Dionysia aretioides* growing in the Amol Valley, N. Iran.
Photo. P. Furse.

Introduction

My interest in the genus *Dionysia* stretches back to 1966 when, as a member of the Wye College (University of London) Southern Zagros Botanical Expedition to Iran I saw my first one, *D. bryoides*, growing on cliffs at Kuh-i-Dinar to the north-west of Shiraz. My interests were fired still further by tempting photographs of various species growing in the wild taken first by Paul Furse and then Jim Archibald. In 1971 I was invited by Professor Tom Hewer to join him on a collecting expedition to Iran and Afghanistan and this enabled me to study a number of *Dionysia* species in the wild more closely.

The genus *Dionysia* is amongst the most tempting and yet tantalising of all genera cultivated by alpine gardeners, containing some of the most beautiful of all cushion-alpines, but at the same time some of the most demanding and difficult to cultivate.

Many gardeners are put off by the very mention of the genus, yet it is true to say that several have proved more amenable to cultivation than was formerly supposed. Their reputation of being impossible to grow and propagate has proved much exaggerated, for nearly all those brought back as seed have been reared to flowering specimens. A number of species, *D. aretioides*, *D. curviflora*, *D. involucrata*, *D. mira* and *D. tapetodes* are now fairly widely grown and well-established in cultivation. At the same time there are others, *D. afghanica*, *D. lamingtonii*, *D. microphylla* and *D. viscidula* for instance, which are far more tricky plants confined to a few specialist collections where they are constantly nursed and pampered.

Besides their beauty (they can be guaranteed to attract attention in the alpine house or on the show bench), it is the challenge of trying to cultivate these species successfully that tempts many alpine gardeners. Certainly this challenge has had its effect, for there are more Dionysias in cultivation today than ever before and our knowledge of their cultivation has vastly improved in the past twenty years.

I first wrote an account of *Dionysia* for the Alpine Garden Society in 1970. In that account 36 species were included. Since then a further 6 species have been discovered in the mountains of Iran and Afghanistan. But for the political strife in the Middle East during the past decade, which has practically brought to a standstill further botanical exploration, other new species would undoubtedly have been discovered.

I have undertaken this new revision of *Dionysia* in an attempt to bring together all the information that has come to light since 1970.

C. Grey-Wilson
July, 1988

I

Brief history of the genus

The genus *Dionysia* was described by Fenzl in 1843 based upon *D. odora*, specimens of which had been collected by Kotschy in Kurdistan some years previously.

Much of the early history of the genus is of generic confusion and generic delimitation. In 1817 Lehmann had described a new species of *Primula*, *P. aretioides*, but in 1846 this was transferred to *Dionysia* by Boissier, who at the same time described three further species of *Dionysia*.

Duby (1844) in his revision of the Primulaceae for the Candolle's *Prodromus* included the genus *Gregoria*, four of the five species of which were later to become species of *Dionysia*. In the same year Duby also described the new genus *Macrosyphonia*, including in it a single species, *M. caespitosa*, which was later to be transferred to *Dionysia* by Boissier.

In 1871 when Bunge published the first monograph of the genus 12 species were known, 5 new ones being included in the work. All these species were included by Boissier (1879) in his *Flora Orientalis*, as they all occurred within the region covered by the flora. Just a few years later, in 1891, Kuntze merged *Dionysia* into *Primula*, but few later authors followed this drastic treatment. Indeed Bornmüller (1899) maintains *Dionysia* and was the first botanist to make important contributions to our knowledge of the genus, basing much of his work on the important collections made by the German, Th. Strauss, who took a special interest in the genus, travelling extensively in W. Persia (Iran) between 1904 and 1908. Despite this, Bornmüller was never wholly convinced by the genus *Dionysia* and many of the species which he included may often be found under alternative names in the genus *Primula*.

In 1905, when Knuth monographed the genus for *Das Pflanzenreich*, the number of species of *Dionysia* had risen to 20, but at this stage no attempt had been made to subdivide the genus.

The exact distinction between *Dionysia* and *Primula* seems to have been a subject of great concern to botanists of the period. In 1909 Pax described a new section within *Primula*, *Dionysiopsis*, including in it a new species *Primula bornmuelleri* and *P. hissarica* (formerly *Dionysia hissarica* Lipsky). Both Pax and Bornmüller believed that *Dionysiopsis* (whether as a section of *Primula*, or genus in its own right) acted as a transition between *Dionysia* and *Primula* section *Floribundae* [sec.]. Smith and Forrest (1928) precipitated matters by excluding the section *Dionysiopsis* from *Primula* altogether, suggesting that they were species of *Dionysia* and it was left to Clay (1937) to make the formal transfer.

Smith and Forrest's conclusions were given added weight by Melchior (1943) who made a detailed investigation of the evolutionary trends within *Dionysia*, including the species of *Primula* section *Dionysiopsis* within the genus. At the same time he became the first person to subdivide *Dionysia*.

Further contributions to the genus were made by Smolianinova (1952) who reviewed the four species coming within the Soviet Union. However, up until the late 1950s very little work was done on the morphological and evolutionary aspects of the genus which had been initiated in 1943 by Melchior.

In 1961 Per Wendelbo published th first modern revision of the genus *Dionysia*, basing his work on detailed cytological, morphological and palynological studies. This work carefully reviews the major difference between *Dionysia* and *Primula*, whilst at the same time discussing the evolutionary trends and migration patterns within the genus. Wendelbo here establishes a well-considered conspectus and subdivisions of the genus, recognising some 28 species in all. This important work has been the basis for all subsequent research and is likely to remain so.

During the 1960s and 1970s Wendelbo contributed a number of important papers on *Dionysia* including an account of the genus in Afghanistan, together with descriptions of 6 new species (1964), an account of the genus for *Flora Iranica* (1965) and several papers, starting in 1976, on the anatomy of *Dionysia*.

During 1970 the present author published an account of the genus for the Alpine Garden Society, basing the classification on that of Wendelbo (1961), but including species described subsequent to 1961, the total number included being 36. Today the genus contains 41 species.

2

Botanical characters

Dionysia can be divided basically into two main types. First there are those species that form lax tufts, somewhat Primula-like, with umbels or superposed whorls (candelabra) of flowers. In contrast there are the majority of species that form neat, often dense cushions made up of numerous leaf-rosettes and bearing solitary scapeless flowers. Between these two extremes almost every intermediate stage can be found, from species with dense cushions and scapes to those with laxer cushions, larger leaf-rosettes and solitary flowers.

HABIT. Dionysias are dwarf shrubs, the branches either laxly or densely arranged. Many form symmetrical, rounded or rather flat cushions, which may attain 1 m diameter in *D. diapensiifolia*, but are generally far smaller. In *D. microphylla*, for instance, mature cushions rarely exceed 15 cm diameter. The branches become woody with age and, in some, the dead leaves fall off leaving the lower stems bare. However, in the majority of species the dead leaves or leaf-bases remain clothing the stem and they are then said to be marcescent.

In species like *D. curviflora*, *D. michauxii*, *D. afghanica* and *D. lamingtonii* the small dead overlapping leaves form closely-packed columns giving the cushion its dense characteristic. These marcescent leaves obviously help to support the cushion, but they may also be fundamental in protecting the base of the plant from extremes of temperature.

In other species such as *D. aretioides*, *D. hedgei*, *D. paradoxa* and *D. teucrioides* the dead leaves or leaf-remains are more spreading, not closely overlapping, and form an altogether looser arrangement.

In all species, though, eventually the stems thicken below and the marcescent leaves are shed leaving a dark rather smooth bark. This may take many years to occur and is less obvious in cultivated plants which are generally relatively young.

The cushion habit also occurs in other genera in the Primulaceae, most notably in *Androsace* where a similar trend from laxly tufted to dense cushion-forming species can be observed.

LEAVES. Leaf shape and size is extremely variable in the genus. Basically two broad divisions can be recognised. In section *Anacamptophyllum* the leaves are revolute with margins folded under. In the larger-leaved species (of subsection *Scaposae*) this characteristic can generally only be observed when the leaves are very young, for as the leaf matures so it becomes flat. This situation is found in species like *D. mira*, *D. paradoxa* and *D. hissarica*. Leaves in these species can be remarkably *Primula*-like, rather thin, toothed and with a network of veins beneath. In subsection *Revolutae* the leaves maintain their revolute character to maturity, the margin folded under so that it lies close to the midrib. In the species of this subsection (*D. aretioides*, *D. revoluta* and *D. archibaldii*) the margin is frequently and neatly toothed. In these species also, leaf dimorphism can be observed; leaves of flowering shoots are revolute as described, however, those of vegetative shoots are produced in the spring and early summer and are often much larger, non-revolute and more like those of subsection *Scaposae* which includes species like *D. mira*, *D. bornmuelleri*, *D. hissarica* and *D. paradoxa*. This indicates a strong affinity between these two subsections.

In the remaining species of the genus, section *Dionysia* and section *Dionysiastrum*, which together include the majority of species, the leaves are small, often rather thick, with a flat or sometimes very slightly involute margin. The leaf-margin in these species is often entire, although in subsection *Caespitosae* (of section *Dionysia*) leaf-dimorphism can often be observed and the leaf-margin is often toothed (as in *D. caespitosa* subsp. *bolivarii* and *D. odora*). In many species in these two sections the leaves are overlapping, often imbricate and only the upper half or third of the leaf is green, the lower overlapped portion being hyaline and without chlorophyll. The veins are generally rather inconspicuous. However, in a few species such as *D. tapetodes*, *D. zagrica*, *D. hedgei* and *D. freitagii* the veins may be distinctly raised and fan-like (flabellate) on one or other surface.

The leaf-surfaces of *Dionysia* are variously embellished with hairs and glands of different sorts. Hairs vary from long articulated structures to shorter gland-tipped ones. This is seen at its most pronounced in *D. lindbergii* where the apical leaf-hairs are very long and articulated, whereas those in the lower half of the leaf are short-stipitate capitate glands. In a number of species (*D. bornmuelleri*, *D. odora* and *D. leucotricha*) the long-articulated hairs are underlain by a layer of short-stipitate capitate glands. The type of hairs and/or glands generally differs from species to species. The hairs may cover the entire leaf or they may be confined to one or other surface, or just to the upper half. Some species are eglandular, whilst in others the entire surface is covered with minute capitate glands, with a complete absence of long-articulated hairs – as in *D. tapetodes* and *D. zagrica*.

Farina is present in a number of species as it is in *Primula*. In *D. hedgei*, *D. involucrata* and *D. microphylla* the farina is of a powdery type very similar to that found in *Primula*. However, in most of the farinose species of *Dionysia* the farina is woolly in nature, consisting of long curved threads of crystals. In several species, most notably *D. tapetodes*, both farinose and efarinose forms are present and it is unclear whether both forms can be found growing in the same colonies in the wild. Farina can be confined to the upper or lower leaf surface, but it is also often found amongst the bracts and calyces, or occasionally on the corolla.

INFLORESCENCE. The inflorescence has been shown (Melchior 1943) to be terminal in the leaf-rosette. Subsequent growth of the rosette is by lateral shoots produced in a leaf-axil close to the base of the inflorescence.

In *Dionysia* the inflorescence ranges from those species in which there is a well developed scape, often, as in *D. mira* and *D. bornmuelleri*, with several superposed verticils of flowers, tiered one above the other as in the candelbra primulas. In other species the scape is progressively reduced, becoming shorter and fewer flowered – *D. caespitosa*, *D. teucrioides* and *D. hissarica* for instance. In *D. lacei* the inflorescence is practically scapeless, but reduced to a cluster of flowers held close against the leaf-rosette.

In the majority of species, however, the inflorescence is reduced to a solitary escapose flower held tight against the leaf-rosette. The fact that there is often a very short scape, often only 1–2 mm long, with one or a pair of inconspicuous bracts, indicates that such a structure truly represents a very much reduced inflorescence

Interestingly the trend in the reduction of the inflorescence from a well-defined scape to a solitary escapose flower can be observed in all the three main sections of *Dionysia.*

BRACTS. In species like *D. mira*, *D. bornmuelleri* and *D. involucrata* the bracts are large and leaf-like and this is true of most of the scapose species. In the majority of species, however, where the flower is solitary, the bracts are often small and inconspicuous, frequently linear or linear-lanceolate. In such species there are generally only 1 or 2 bracts per flower.

The indumentum details of bracts resembles that of the leaves.

CALYX. In *Dionysia* the calyx is campanulate or tubular-campanulate with 5 lobes or teeth. The separation of the lobes is very variable from one species to another – in some species the calyx is split for only a quarter or a third of its length, whereas in others the calyx is split for three-quarters of its length or practically to the base.

As with the bracts, the indumentum details of the calyx closely match those of the leaves.

COROLLA. The corolla can be yellow, pink or violet depending on the species, but no species has both yellow and pink or violet forms as far as is known. The corolla is salver-shaped with a long slender tube, generally much exceeding the calyx, and a broad limb consisting of 5 lobes.

As in *Primula* the corolla is heterostylous with both pin- and thrum-eyed forms. This occurs in all *Dionysia* species except *D. involucrata* and possibly *D. teucrioides*. Pin- or thrum-eyed forms always occur on separate plants. In the former the stamens are housed between on third and two thirds the way down the corolla-tube and the style reaches beyond to the throat or is exserted from the mouth of the corolla. In the later (thrum-eyed) the situation is reversed with the stamens being housed in the throat of the corolla whilst the style reaches between one third and two thirds along the corolla-tube.

Heterostyly ensures cross-pollination, for an insect transferring pollen from one individual to another will only transfer pollen from a pin- to a thrum-eyed flower, or vice versa, because of the relative positions of stamens and styles in the flowers. This is precisely the situation in the common Primrose, *Primula vulgaris*.

Even if pollen inadvertently reaches the stigma of the same flower it is very unlikely to lead to fertilisation for another biological device prevents it doing so. Within a species of *Dionysia* the pollen grains of pin- and thrum-eyed flowers are a different size. Furthermore the stigmatic surface consists of different size papillae. Accordingly, the corresponding pollen grain must land on the stigmatic surface (ie pollen from a pin-eyed flower on the stigma of a thrum-eyed plant, or vice versa) or it will fail to germinate, or even if it does, will rarely lead to subsequent fertilisation.

This dual incompatibility system ensures cross-pollination. It has implications for the grower who must possess at least one pin-eyed and one thrum-eyed plant of a particular species to ensure fertilisation and the production of seeds.

Even then the grower will almost certainly have to hand-pollinate to ensure success.

The corolla-lobes vary a great deal in size from one species to another and may be entire to emarginate or sometimes bifid. The corolla may be glabrous outside, pubescent or glandular, again depending on the particular species.

FRUIT-CAPSULE. The fruit-capsule is rather uniform in all species of *Dionysia*. At maturity the small egg-shaped capsule is pale brown, and splits into five valves. Seed number per capsule varies from species to species. However, scapose species such as *D. mira* and *D. bornmuelleri* have a large number of seeds per capsule, whereas the tight caespitose species with solitary flowers have a reduced number of seeds per capsule – *D. michauxii* and *D. lamingtonii* for instance have only 2–3 seeds per capsule.

Seeds are relatively larger when there are only a few per capsule, so that species with numerous seeds per capsule have, in consequence, the smallest seeds.

There is some evidence (Wendelbo 1961) that in at least some of the densely caespitose species of *Dionysia* the seed capsule are autochorous, bursting suddenly to expel the seeds forcibly from the capsule. As most of these species are chasmophytes (cliff-dwellers) then such a system would help fling the seeds into suitable cliff crevices. Wind may help in such dispersal.

12 *Dionysia aretioides* planted out in the new alpine house, RBG Kew. Photo. B. Mathew.

13 *Dionysia aretioides*– a 10 year old plant in cultivation. Photo. E. G. Watson.

14 Limestone gorge, Kuh-i-Sabzpuchon, S. Iran. On the cliff walls grow *D. diapensiifolia*, *D. bryoides* and *D. revoluta* subsp. *revoluta*. Photo. C. Grey-Wilson.

15 *Dionysia revoluta* subsp. *revoluta* on the limestone cliffs of Kuh-i-Sabzpuchon, S. Iran. Photo. T. F. Hewer.

3
Cytology

The cytology of *Dionysia* remains in its infancy and a good deal more needs to be researched before any positive views can be put forward.

To date 6 species have been investigated, *D. aretioides*, *D. bornmuelleri*, *D. hissarica*, *D. involucrata*, *D. revoluta* and *D. tapetodes*. All these have been shown to have $2n = 20$ chromosomes. Unfortunately all but *D. involucrata* belong to section *Anacamptophyllum* and no member of the largest section, *Dionysia*, has been investigated. However, as so many species are now in cultivation (about half the genus) then it should be possible to do further investigation, providing suitable root material can be obtained.

In contrast, in *Primula* section *Sphondylia* (= *Floribundae*), those species that have been investigated (*P. floribunda*, *P. gaubaeana* and *P. verticillata*) have been shown to have $2n = 18$ chromosomes. In this section, then, the basic number of chromosomes is $\times = 9$, whereas in *Dionysia* it would appear to be $\times = 10$.

A basic number of $\times = 10$ is also found in *Androsace*, *Vitaliana*, *Hottonia* and *Soldanella*. It is also a characteristic of *Primula* sections *Muscarioides* and *Soldanelloides* (see Bruun 1932, Favarger 1958 and Smolianinova & Kamelina 1972). Despite a similar chromosome number these later two *Primula* sections have nothing to do with *Dionysia*.

4
Relationship to Primula

Dionysia and *Primula* are clearly very closely related, indeed many find it difficult to see why the two genera are kept separate. Unfortunately there is no single character that satisfactorily divides the two genera and a combination of characters must be employed. These are listed below:

1. WOODY STEMS. All *Dionysia* species develop woody stems with age, in some such as *D. bryoides* these may reach as much as 2 cm diameter. In most species the stems are covered during the first few years by marcescent leaves and leaf-remains. Eventually, however, the cortex and dead leaves are shed, leaving a dark brown or bluish-brown bark behind. Woody stems are also found in some species of *Androsace*, in *Douglasia* and in several sections of *Primula* (*Bullatae* and *Dryadifolia* etc).

2. LEAF-MARGIN. In *Dionysia* the leaf-margin is revolute or flat or slightly incurved, but never involute. In *Primula* the vernation of the young leaves has always been considered of major importance and the species can be divided into the *Involutae* and the *Revolutae*. *Dionysia* shows most morphological similarities (especially section *Anacamptophyllum*) with *Primula* section *Sphondylia* (= *Floribundae*): however, this section is characterised by its leaves being clearly involute in the young state and the Primula-like species of *Dionysia*, such as *D. mira*, *D. bornmuelleri* and *D. paradoxa* can always be separated by this character from *Primula* species of section *Sphondylia* – *P. verticillata* and *P. floribunda* for example (Plates 1 and 2).

3. COROLLA-TUBE. The long narrow slender corolla-tube is a feature of most species of *Dionysia*, generally being 3–4 times the length of the calyx. This distinguishes *Dionysia* from many species of *Primula*, though this character cannot be used *per se* to distinguish the two genera.

4. FRUIT-CAPSULE. The small subglobose capsule of *Dionysia* splits into 5 valves. In *Primula*, in contrast, seed capsules mostly dehisce at the apex into 5 or 10 teeth. Only in *Primula* section *Bullatae* does a 5-valved capsule occur. Several phyletic lines can be observed in Primulaceae for 5-valved capsules also occur in *Androsace*, *Douglasia* and *Vitaliana*.

5. SEEDS. There is a clear trend in the reduction of seed numbers in *Dionysia*. Although some species have numerous seeds per capsule, as in *D. bornmuelleri* and *D. mira* and most of the species of subsection *Scaposae*, most of the densely caespitose species have very few seeds per capsule, often only 1–3. A few-seeded capsule is not known in *Primula*.

6. FARINA. Farina is present in many *Dionysia* species, but is absent from most species of subsection *Bryomorphae*. In all species of *Dionysia*, with the exception of *D. microphylla* and *D. hedgei* and *D. involucrata*, which have powdery type farina, the farina is of the woolly type. All farinose species of *Primula* possess farina of the powdery type. The structure of hairs is very similar in both *Primula* and *Dionysia*, but when present the farina type clearly distinguishes these two genera.

7. POLLEN. Dionysias generally have many furrowed pollen grains – with 4–10 (often 6–8) furrows. In *Primula* the pollen grains generally have 3–4 furrows, but it is important to note that in the morphological similar section *Sphondylia* the pollen grains are always 3-furrowed. In *Primula* section *Vernales* the pollen grains are rather similar to those 4-furrowed ones sometimes seen in *Dionysia*, but the plants bear little morphological similarities otherwise.
8. CHROMOSOME NUMBER. Few species of *Dionysia* have had their chromosome numbers counted. Those that have reveal a chromosome number of $2n = 20$ ($\times = 10$). In *Primula* the chromosome number is variable; $\times = 10$, the basic number, is found in sections *Muscarioides* and *Soldanelloides*. In section *Sphondylia*, however, the basic number is $\times = 9$. $\times = 10$ is also the base number for both *Androsace* and *Vitaliana*, whilst it is also found in *Hottonia* and *Soldanella*.

From this accumulated evidence it is clear that although *Dionysia* and *Primula* are more closely related than *Dionysia* is to any other genus in the *Primulaceae*, the two genera can be separated on a number of characters. Although no single character can be used in isolation to separate *Dionysia* and *Primula*, this unique combination of characters isolates *Dionysia*. For instance woolly-type farina and a few-seeded fruit-capsule, although not present in all species of *Dionysia*, never occur in *Primula*. It could of course be argued that *Dionysia* would be better included in *Primula*, but then it would most certainly have to be granted subgeneric status and at this level it matters little – certainly nothing would be gained from such a move.

5
Evolutionary trends in Dionysia

Wendelbo (1961) has very adequately summarised the evolutional and migrational trends in *Dionysia*, based in large measure on earlier work done by Melchior (1943). Melchior, however, based his analysis on rather limited material – since then more material has been collected and new species have been discovered. Melchior showed that within *Dionysia* various evolutionary trends ('progressive trends') could be clearly seen with the most highly evolved species being those which are more reduced in their characters. What is most interesting is that these 'progressive trends' occur in all the three major sections of the genus. Melchior's scheme, modified and elaborated by Wendelbo, can be summarised as follows:

1. INFLORESCENCE. Scapose with a number of superposed verticils of flowers. to a simple umbel, to a one-flowered short scape, to a solitary sessile flower.

2. BRACTS. Numerous foliaceous bracts, to few ovate, entire bracts, to two lanceolate or linear bracts, to a solitary linear bract.

3. VERNATION OF YOUNG LEAVES. From revolute to flat.

4. LEAF-LAMINA. Thin, large and petiolate, to smaller.

5. LEAF-MARGIN. Toothed and bidentate or bicrenate, to simple-toothed, to obscurely toothed, to entire.

6. LEAF-LAMINA VEINS. Veins reticulate and often raised beneath, to pinnately veined, to fan-like, to a simple mid-vein with few branches.

7. SEEDS PER CAPSULE. Many (20–30) to few (1–3).

8. HABIT. Loose tufts, to dense cushions, to small compact cushions.

9. COROLLA COLOUR. Yellow, to pink or violet.

10. CALYX. Split for only part of length, to split to the base.

11. COROLLA-LOBES. Deeply to shallowly emarginate, to entire.

12. FARINA. Present, to absent.

6

Cultivation

Dionysias present a challenge to the grower. Of the 20 species in cultivation only 5 can be said to be reasonably easy to grow – *D. aretioides*, *D. curviflora*, *D. involucrata*, *D. mira* and *D. tapetodes*. The remainder are difficult or very difficult to grow.

Those venturing into *Dionysia* cultivation for the first time would be well advised to start with the five species named above – perhaps the easiest of all are *D. aretioides* and *D. curviflora*. The great challenge of *Dionysia* growing is not just growing perfect floriferous cushions for the show-bench (that is only of secondary importance), but understanding the needs of the different species and, more importantly, discovering how to propagate them.

Propagation is the key to success. Henrik Zetterlund writes 'The key to success with *Dionysia* is propagation and the rare species are rare because they are b...ds to propagatioin.' It is often not possible to propagate from seed as only relatively few species produce this regularly and, except for *D. involucrata* and possible *D. teucrioides*, seed will only be possible if one has both pin-eyed and thrum-eyed plants of the same species in ones collection. At the present time no wild-collected seed is being introduced and this situation is likely to remain the same for some time to come. Cuttings are therefore the only option – as with any rare or difficult species, serious growers will want to have more than one specimen of each species in a collection to ensure continuation. Mastering propagation will ensure this, whilst at the same time surplus young plants can be given or exchanged with other growers.

General requirements. In the wild, Dionysias mostly inhabit shaded or partially shaded cliffs and rocks. The summers are hot and extremely dry, the winters generally intensely cold with most precipitation falling as snow or early spring rains. The roots of Dionysias, like those of many chasmophytic plants, delve deep into rock crevices to seek out precious moisture. In the spring the cliffs may be wet and water may seep across the cliffs and around the plants. This is the time when active growth commences and the species come into flower. As summer approaches the rocks generally become very dry – at least on the surface. Shaded or partially shaded cliffs present a more equable regime to which Dionysias have become well adapted. Full sun would present a harsher regime – greater fluctuations in temperature between day and night as well as more severe aridity – so it is not surprising that few Dionysias are found in such aspects.

Conditions in Europe and other temperate regions are clearly not like those in the mountains of Iran and Afghanistan. The climate is cooler and very much wetter, indeed rain can fall at almost any time of the year. This naturally affects plants, even if they are sheltered within the confines of a frame or alpine house. Careful watering and ample ventilation, together with the correct choice of compost are key factors, but there is no simple recipe for success and each sucessful grower has his or her own particular set of rules. In the following pages I have tried to present these varying opinions and the reader will need to choose those which suit his or her own particular conditions and growing methods.

It is important to understand the cycle of growth of a typical Dionysia in cultivation and I cannot describe this better than by quoting from a letter sent by Henrik Zetterlund – 'There are two extremely critical periods in the Dionysia season. The first is late March-early April as the sunshine is becoming more intense, when the plants require slight shading in order not to be burnt on the south-facing part of the cushion – later shading is not necessary providing watering is sufficient. The other critical period is early autumn (September-October) when the plants seem particularly susceptible to botrytis ... Like so many other 'ever-active' plants from the Dionysia-belt, they have two periods of growth each year with a short summer dormancy in between. During this dormancy cuttings strike poorly and feeding can also be withdrawn for obvious reasons. During the second period of growth they often produce some flowers which we (in Göteborg) invariably remove to decrease the botrytis attacks ..."

COMPOSTS. Nowhere is Dionysia cultivation more controversial than that of compost mix and components. All will, however, agree that the most critical criterion is that the compost be extremely well-drained – a high proportion of coarse grit therefore features in every compost.

The simplest compost consists of 2 parts John Innes Potting Compost No. 2 or 3 and 1 part coarse grit or limestone chippings and a number of growers have had great success with such a mix.

Henrik Zetterlund writes that at Göteborg 'The compost we are using consists of 1 part neutralized sphagnum-peat, 1 part coarse sand, 1 part sterilized loam and 3 parts granite grit (2–5 mm), a rather simple mixture that can be compared with John Innes mixed with grit'.

Chris Norton recommends – 'a very gritty mix – using Croxden or Shap grit plus crushed tufa, crushed pot or brick and fine Perlag with perhaps 30% mixed leaf-mould and John Innes No. 3. I'm not particular about the JI – I often use a home mix with a Chempak base formula'.

Michael Kammerlander at the Würzburg Botanic Garden, another skilful cultivator of alpines, grows Dionysias in a compost of roughly equal parts of coarse sand, perlite, vermiculite and betonite.

The compost mixture depends to some extent on growing methods. If pots are plunged in a bed of sand or gravel then the more free-draining the compost the better. However, if pots are free-standing then a slightly more moisture retentive compost is advisable as the pots will tend to dry out quicker – the balance can be made from the existing formulae by simply reducing the percentage of grit.

Composts should not be too rich. Too rich a compost will result in soft lush growth and such plants may look 'out of character' and, furthermore, they tend to be short-lived. The harder the plants are grown the better, though this does not mean that they should be starved of all nutrients.

POTS. Most growers favour the use of clay pots and certainly if pots are plunged on a bench in the alpine house then clay ones are easier to control – it is easy to over-water in plastic pots and they can take considerably longer to dry out.

Pots should have ample crocks at the base or a deep layer of coarse gravel.

POTTING-ON. Young plants and more mature ones that have filled their pots will require potting-on, undoubtedly one of the most tricky operations when losses can be expected, even in the hands of expert Dionysia growers.

The best time to undertake this task is when active growth has commenced in the spring, or shortly after plants have finished flowering. Plants should be watered beforehand by plunging pots to their rims in a basin of water – this will help ensure that the compost and roots stay knit together and don't fall apart once removed from the original pot. The less plants are repotted the better, so it is wise to choose a new pot large enough to contain the plant for at least the next 2–3 years, if possible.

It is very important to ensure that plants are repotted at the same level as they were previously, if anything leaving them very slightly proud. On the other hand, potting them too deep will prove disastrous.

The compost should be firmed in very well around the plant and the top 1–2 cm finished off with a layer of coarse grit, working it in carefully under the plant. Most growers in fact wedge small pieces of rock or tufa around the neck of the plant. This helps to keep this vulnerable part of the plant free of excessive moisture. Furthermore a number of growers maintain that plants grow better wedged on a collar of rock.

After potting-on, providing the weather is warm, then the plants can be plunged in a basin of water to wet the new compost adequately, but they must not be left there too long.

It is important to remember that plants must be handled with the greatest of care when attempting to repot them. Any bruising, especially of the tight cushion-forming species, may result in die-back or fungal infection – especially botrytis. Henrik Zetterlund writes – 'Repotting of adult specimens is carried out in May–June (at Göteborg). We use the usual compost (see p. 30) and topdress with 2–3 cm of chippings. A method that in some cases inhibits cushion-rot is to place the plant so that the cushion is situated a centimetre above the pot-rim and build up around the plant with pieces of rock. After repotting, the pots are thoroughly watered and kept slightly shaded for two weeks. Like most chasmophytes, Dionysias detest over-planting so you can't get a sprawling specimen into shape by planting it deeply in a large pot...'

Large plants of the laxer growing types such as *D. mira* and *D. aretioides* can be split apart carefully into several pieces and each treated as a separate plant. However, none of the dense cushion-forming species can be trifled with in such a manner for they will assuredly die within days, if not hours – see under propagation, p. 41.

TUFA. When many of the Dionysias came back into cultivation during the 1960s and 1970s a lot of growers tried growing young plants on tufa blocks, and some still do to this day, although this method now meets with less approval than formerly. Plants inserted in holes in tufa grow more slowly than the equivalent plant in compost. However, they can often keep a good tight habit, but seldom reach a large size as they are eventually inhibited by the tufa and the size of the root cavity – they cannot be moved on once established as they could in a pot of compost! There is no doubt though that plants can look very appealing on a block of tufa and most growers have one or two plants so placed. For large collections tufa-growing is impractical and time-consuming.

WATERING. It is often put about that Dionysias require little and very careful watering. Although the latter statement is undoubtedly true the former is totally misleading.

The watering of Dionysias is a complicated procedure. Unfortunately it is true to say that only with experience can one learn the requirements of individual species. It is possible, however, to outline the basic requirements.

Stan Taylor writes – 'Not many Dionysias die from a lack of water, but a lot die from a surplus of it'. However, that is not to say that during the active growing season watering should be restrained, indeed plants can take in a good deal of water at such a time. The careful control of watering at different times of the year is critical.

During the autumn and winter is the most critical time, for excess water then, even if it does not harm the plant directly, may encourage botrytis infection. Pots in a sand plunge are the easiest to handle what ever the time of the year. During the autumn and winter most growers withhold water from the pots themselves and water only the surrounding sand. This is then only rewatered when it is seen to be drying out and certainly never when the weather is cold and damp.

As the spring arrives and growth commences then watering can be increased, but a careful watch must be kept on plants at all times to ensure that they are not overwatered. During the peak growing season, in late spring and summer, plants should be kept well watered and during this time the pots can be watered directly. Some growers in fact advocate removing pots from their plunge and soaking them in a basin or bucket of water with the pots up to their rims. Again this is best done only when the weather is warm, or hot and dry.

Experience will tell when a plant requires water. Henrik Zetterlund remarks that – 'During the winter, watering is limited to soaking the plunge when it's starting to dry out. In summer the plunge is kept rather moist and during sunny periods the pots are watered once a week with a liquid fertilizer, 1.5 promille solution (3.0 for the large species). Larger plants may need water more often and this will have to be learnt from experience, best by caressing the cushions gently to judge their turgor'.

A sure sign of overwatering is when leaves all over the plant begin to turn yellow and parts of the plant die or the whole plant collapses (see also pests and diseases, p. 41)

Species such as *D. aretioides*, *D. mira* and *D. curviflora* will tolerate overhead watering from time to time. However this should only be attempted on dry days when the plants are in active growth and well before nightfall. Plants should be shaded from direct strong sunshine if watered overhead and they should have dried out thoroughly before night – cool moist conditions may invite fungal infection, especially on the dead leaves. The soft, hairy, tight-cushioned species such as *D. michauxii* and *D. lamingtonii* will not tolerate overhead watering and will die back in patches or overall very rapidly, even if the cushion is only slightly and accidently watered.

SHADING. During the late winter and early spring (March and early April) when growth commences, light shading will prevent scorch of the young growth, especially on the side of the cushion receiving the most sun.

In general, shading is not required in the summer, providing the water regime

16 *Dionysia revoluta* subsp. *canescens*.
Photo. S. Taylor.

17 *Dionysia archibaldii*.
Photo. S. Taylor.

18 *Dionysia caespitosa* photographed near Esfahan, Iran.
Photo. T. F. Hewer.

19 *Dionysia diapensiifolia* growing on the limestone cliffs of Kuh-i-Barfi, S. Iran. Photo. J. C. Archibald.

20 *Dionysia diapensiifolia* growing on the limestone cliffs of Kuh-i-Barfi, S. Iran. Photo. J. C. Archibald.

21 Cliffs of conglomerate, Kuh-i-Dinar, S. Iran, the type locality of *D. termeana* (some plants can be seen beneath the overhang – left). *D. bryoides* also grows on the same cliffs. Photo. T. F. Hewer.

22 *Dionysia termeana* growing on the conglomerate cliffs of the Kuh-i-Dinar, S. Iran. Photo. T. F. Hewer.

24 *Dionysia lamingtonii* growing on limestone cliffs, Zagros Mts., W. of Shahreza, W. Iran. Photo. T. F. Hewer.

23 *Dionysia haussknechtii* growing on limestone cliffs S of Aligudarz, W. Iran. Photo. J. C. Archibald.

is correct and there is plenty of ventilation in the alpine house or frame to keep the atmosphere buoyant and prevent temperatures reaching too high. A fan blower may be of great help, especially at those times of the year – particularly the early spring and autumn – when botrytis infections can be expected. Despite this, most growers provide some shading during the summer months. Remember that most species inhabit shaded cliffs in the wild; cushions may scorch during hot sunny weather. Temperatures in excess of 30°C during hot summer spells may adversely affect plants.

Even during the winter, except during periods of severe cold, houses should be generously ventilated.

HARDINESS. Dionysias grow in regions subjected to intensely cold winters. In cultivation they are completely hardy. Geoff Rollinson writes – 'In the winter of 1981–82 I recorded −25°C in the alpine house and *all* plants including Dionysias were frozen for 2–3 weeks. My only losses were immature rooted cuttings which I now wean through their first winter using a heating cable in the sand plunge'.

Plants are more likely to survive such low temperatures if they are reasonably dry at the time.

In Göteborg, Henrik Zetterlund tells me that the Dionysias are in fact kept in a frost-free house with a winter minimum temperature of +1°C.

GENERAL CARE. Dionysias need to be kept clean at all times. On the larger species such as *D. mira*, dead leaves should be carefully removed as they may prove to be sites of fungal infection or harbour various pests such as aphids and red spider mites.

Dead flowers (corollas in particular) may also be sites for fungal infection. Scapose species are easily cleaned of dead flowers by removing the entire inflorescence (providing they are not wanted for seed production) after flowering. Cushion species are more tricky and delicate; the dead corollas need to be picked off carefully using a pair of tweezers. Removing corollas in this way does not harm any potential seed production.

Similarly, dead branches or parts of cushions should be removed. Damaged parts of cushions should be treated with a fungicide in powder form to prevent fungal infection.

As cushions expand, more pieces of rock can be wedged carefully under the plant, especially the laxer cushions of species like *D. aretioides*. This helps to keep cushions more compact and saves them from falling apart at the centre.

PROPAGATION. The propagation of Dionysias presents a real challenge to the grower. The difficulty of propagating many species is primarily the reason for so many being scarce in cultivation. Very few growers can be said to have mastered the propagation of this difficult group, and those that have would be most loathe to admit it.

One should not give the impression that all the species are of equal difficulty when it comes to propagation, for the handful of species widely grown are relatively simple in this respect – particularly *D. aretioides, D. curviflora, D. involucrata, D. mira* and *D. tapetodes*. Unfortunately some of the most beautiful, gems such as *D. afghanica, D. lamingtonii, D. microphylla* and *D. viscidula* have proved, over the past 20 years, difficult to propagate in any quantity.

Much of the problem lies in the fact that seed of these rarer species is never produced in cultivation. This is because many of them exist as a single clone, either pin- or thrum-eyed, and without both in a collection there is little or no chance of any seed.

Where both types exist in a collection it is wise to hand cross-pollinate plants to try and ensure some fruit-set. The developing fruits must then be watched with great care as they mature so that the seed can be gathered and not lost. If pin- and thrum-eyed plants exist in different collections then there is a strong case for loaning plants to others or at least using them for cross-pollination purposes.

SEED. Like that of *Primula*, *Dionysia* seed is shortly viable and should be sown the moment it is ripe, or as soon after as possible. Seed sometimes germinates almost at once, but may not do so until the following spring. There have been several reported cases of seedlings continuing to appear two or three years after sowing, so one should never be too eager to discard seed pans.

Seed of the scapose species is obviously easy to collect, both in the wild and in cultivation. The ripe fruits can be removed and crushed gently to release the seeds into a suitable container. Seed of the cushion-species is more difficult as the fruits are lodged well down in the leaf rosettes. When seed is collected in the wild it is the usual practice to dry chunks of cushion. When these are thoroughly dry they are crushed and the debris, including seeds with luck, is sown as if it were all seed. This method has proved highly successful in the past. Growers, of course, will not want to sacrifice cushions or parts of cushions in such a way in the hope that there might be seed in them somewhere. A close inspection of cushions in the summer should reveal fruit capsules nestling in the middle of leaf-rosettes. As these burst they reveal small 5-pointed stars (the 5-parted capsule walls). In some instances seed is flung from the plant so it will probably be lost. However, capsules can be removed with a fine pair of forceps – this can be both fiddly and time consuming, but well worth the effort if more plants can be raised. Seed of many cushion plants needs to be collected in a similar way.

Seed is fairly readily obtainable from the following species: *D. aretioides*, *D. curviflora*, *D. freitagii*, *D. involucrata*, *D. mira* and *D. tapetodes* and occasionally from *D. archibaldii*, *D. microphylla*, and *D. teucrioides*.

There should be a reasonable chance of producing seed from hybrid crosses and it has been shown for instance that crosses within section *Anacamptophyllum* produce some interesting and floriferous plants, easily raised from F1 generation seed – the cross between *D. aretioides* and *D. teucrioides* shows particular promise. Little has been done to cross caespitose species in subsection *Bryomorphae*. The species may not cross easily of course, but the potential from such crosses is enormous, particularly as regards vigour, ease of propagation and colour range. If such crosses are produced then great care must be taken to safeguard the integrity of the species which may easily be swamped by more easily grown hybrid offspring.

The seed compost should be fine but well-drained and should be firmed down and soaked before sowing: a suitable compost is John Innes Seed Compost with a third extra grit added, or similar. Seed needs to be sown as thinly as possible and covered with the finest layer of silver sand. Pots can be placed in a cold frame or glasshouse, with or without a piece of glass placed on top of the pot. Care must be taken to ensure that the compost does not dry out, otherwise any germinating seeds may perish.

Most species require a cold period and for that reason most seeds seem to germinate in spring. Harold Esselmont wrote in 1969 – 'I have found raising Dionysias from seed or brushings from cushions a gamble in most cases, except for *D. aretioides*, where one may expect a reasonable germination. I sow mine in trays in December, covering them with a layer of chippings, and snow when available, and try to ensure that they get a good freezing'. Peter Edwards wrote at the same time – 'Propagation is straightforward from seed, if available, but certain of the species will not germinate in the same year or for a number of subsequent years for that matter'. Harold Esselmont again – 'A pan of *D. aretioides* sown in August 1966 and allowed to dry out in the autumn months and rewatered in November, has just (April 1969) produced half a dozen seedlings – after almost 3 years'.

These comments and others that I have since received call into doubt the short viability of *Dionysia* seed. It is obviously desirable to sow seed as soon as possible after it ripens, but seed is capable of germinating several years after sowing. It may be that it requires a good deal of freezing before germination can take place, although many growers find that a minimum of 5°C is sufficient for most species. The fact that some species appear to germinate spasmodically over a number of years may be a built-in safeguard in Dionysia evolution to ensure survival of seed in unfavourable years – an important factor for plants that have such exacting habitat requirements.

'Growing on' seedlings presents a major problem and it is debatable whether seedlings should be potted on at a young stage or left to develop for a while, and the choice depends on the grower and on the season to some extent. Obviously it would be foolhardy to pot on seedlings in the autumn when growth is slowing down. Harold Esselmont prefers the early approach – 'When seeds germinate in the spring, I transfer them at an early stage, before a root system has formed, to thumb pots, using a gritty tufa mixture and wedging the neck of the plants very carefully between two small pieces of soft tufa. The pots are bottom-watered and plunged in sand in a shady north frame'.

When seedlings are well spaced in a pan then there is less urgency to pot them on quickly. However, it is important to do so very early if 'lumps' of Dionysia cushions have been sown, for as Peter Edwards pointed out (1969) – 'Seed from deep in the cushion germinated, but care in getting them out of the cushion was necessary to avoid breaking or damaging the roots which would mean certain death to the plant when potting-on'

Young Dionysia seedlings often grow vigorously, generally producing uncharacteristic foliage to begin with before they settle down into a slower regime. Once seedlings are established they can be given more light and transferred to the alpine house and treated as are the adult plants. The first winter is often critical and some protection from heavy frost may help the survival rate. As far as the young plants are concerned the fewer times they are potted on the better – especially the rare caespitose species.

CUTTINGS. The main means of increasing those species of Dionysias in cultivation at the present time is by cuttings. Some species like *D. aretioides* and

D. curviflora root readily from cuttings but the majority are far more difficult.

Patience is the keyword in propagation from cuttings; Dionysia cuttings are small and awkward to handle and take some time to root.

The normal method is to take single-rosette cuttings, removing them from the parent plant with a pair of fine scissors or a razor-blade. Methods vary from one grower to another. Henrik Zetterlund in Göteborg describes his methods as follows – 'Normally we take one-rosette cuttings with a stem as long as possible, clean it, cut the stem with a razor-blade, dip it in rooting hormone and insert it in a mix of sand and water-retentive rockwool.* Since most of the cuttings are inserted in late May care has to be taken that they aren't subjected to too much heat, so to keep the temperature even at about 20°C we move the tray to a cooler spot in warm weather. Rooting may take from 3 weeks to 6 months. A cutting that has rooted is clearly distinguishable from those that haven't and should be planted in compost as soon as possible – they urgently need more nutrients to get going. We grow the young cuttings in 5 cm plastic pots for one year before they get "adult treatment"'.

Geoff Rollinson remarks – 'Cuttings are taken, where possible, from new emergent growth after flowering, usually in late May and early June – rooting normally takes place in 4–6 weeks'. Cuttings are placed in plastic pots with the lower third of the pot filled with moist peat, the upper two thirds with fine silver sand. These pots, with the cuttings inserted, are stood on seed trays (without holes) with a layer of moist peat and placed in a cool, light but not too sunny position in the alpine house or cold frame.

Stan Taylor also described his method of taking cuttings to me – 'I use the same method for all Dionysia cuttings, but the success rate varies quite considerably. A three or four inch (7.5 or 10 cm) plastic pot is used with a plastic dome. The pot is filled with sharp silver sand and it is important that the sand is sharp and not rounded. This is then watered until saturated and left to drain. The cuttings are all taken as single rosettes with as much stem as possible, removing the stem leaves with a razor-blade and dipping them into a rooting hormone. They are put into the pots, ensuring good contact between the cutting and the sand. The pots containing the cuttings are placed in a cool shady place until rooting takes place. Pots *must not* be watered again otherwise damping-off will occur. Rooted cuttings should be potted up as soon as possible – I would suggest as soon as the roots are 4-5 mm long. After potting-up the cuttings are plunged in sand in a shady place and kept moist until established.'

Chris Norton comments that 'I always have cuttings on the go and take them at anytime. I do them in batches in 2 in (5 cm) pots of coarse sand, plus peat and pumice (mostly sand), keeping the pots in a propagating case. When there are signs of rooting I float the cuttings out of the sand to avoid root damage. Using a very gritty mix I find I can safely 'overpot' and this gives 2 or 3 years growth without disturbance'.

For other remarks on propagation see under the individual species.

Once cuttings are rooted the next difficult stage is to establish them in pots in

* Rockwool consists of melted basalt rock formed into inert, sterile strands and (often) treated with a water-repellant coating; the chief advantage lies in its behaviour when overwatered, since the trapped air is not displaced and remains available to the root system.

a suitable compost, see p. 30. The sooner after rooting this is done the better, as it is important that cuttings 'get away' quickly and are not 'checked'. Losses at this stage can be expected but with care and experience this can be kept to a minimum. Getting the young cuttings through their first winter can also be a problem and care must be taken not to overwater them during this time and to fend off fungal attacks – especially on caespitose species with very hairy rosettes such as *D. michauxii*.

DIVISION. Division is not a method that can be successfully used on many *Dionysia* species. However, several of the coarser ones can be treated in such a manner – members of subsection *Scaposae*, for instance, or *D. aretioides* and *D. revoluta*. The parent plant should be divided carefully to minimize bruising and each portion should have its fair share of roots. The portions can then be potted up singly, kept moist and lightly shaded until they are establised, then treated in the normal way.

PESTS AND DISEASES. Dionysias do not suffer from many pests or diseases. Aphids (both root and stem) can be a nuisance, especially in the early spring, and should be watched out for. Signs of attack may be indicated by yellowing of the foliage (this may also indicate overwatering). If aphids are discovered treatment with a proprietary insecticide will usually quickly solve the problem.

The worst disease that can afflict Dionysias is undoubtedly botrytis. This may vary from a localized infection to a more general one. Botrytis may attack weak or unhealthy plants, but it frequently infects dead tissues to begin with – dead or dying leaves and dead flower remains (especially on caespitose species). If the plants are kept 'clean', removing dead tissues carefully then this will help to minimise possible infection. The autumn, winter and early spring are the most vulnerable times of the year, especially when the weather is humid and overcast. Houses should be very well ventilated at all times except during severe cold periods. The use of a fan blower to circulate the air will also be a valuable aid and help reduce attacks of botyrtis. Once an infection has set in then all infected parts should be removed and the plant treated with fungicide to prevent re-invasion – Benlate is suitable.

Some preventative measures can be taken in the autumn to reduce fungal attacks and several growers advocate giving plants a soaking of Benlate, by bottom watering, as a general practice at this time of the year.

7
Taxonomic Treatment

Dionysia

Dionysia Fenzl, Plantarum Generum et Specierum Novarum. Decas Prima. Flora 26: 389 (1843). Type: *D. odora*.

Gregoria Duby, Primulaceae in DC., Prodr. 8: 45 (1844), *pro parte*.

Macrosyphonia Duby in Mem. Soc. Phys. Hist. Nat. 10: 426.

Primula sect. *Dionysia* Kuntze in Post & Kuntze, Lexicon Generum Phanerogamarum, Stuttgart: 406 (1904).

DESCRIPTION. *Dwarf shrubs* or suffrutescent perennials, forming lax to dense tufts, often compact and caespitose, usually aromatic. *Stems* moderately to much-branched, usually becoming thick and woody below in older plants, covered for a part or the whole of their length in spreading or imbricate, marcescent leaves or leaf-bases. *Leaves* spirally arranged, evenly spaced or crowded into whorls, but stems always terminating in a lax or congested leaf-rosette, revolute, flat or very slightly involute in bud, sometimes revolute at maturity otherwise flat, varying greatly in size from large (7–8 cm long) to small and scale-like (2–4 mm long), with or without a petiole, the margin toothed to entire, the veins sometimes raised, and flabellate (fan-shaped), on one or both surfaces, covered on both or either surface with articulate hairs, sessile or stipitate glands or a mixture of both, rarely subglabrous, farinose or efarinose, the farina when present white or yellowish and woolly, rarely powdery. *Inflorescence* terminal, scapose, bearing a simple umbel or 2–7 superposed whorls (verticils) of flowers, or reduced to a single escapose, sessile or short-pedicelled, flower. *Bracts* foliaceous to linear or lanceolate, toothed to entire, generally pubescent and glandular like the leaves. *Calyx* tubular to campanulate, divided for half or more of its length into 5, entire or occasionally toothed, lobes, usually pubescent and glandular like the leaves. *Corolla* heteromorphic, rarely homomorphic, yellow, pink or violet, hypocrateriform (salver-shaped), exannulate; tube slender, at least 3 times the length of the calyx, swollen at the insertion of the stamens; limb flat, divided into 5 elliptical, ovate to suborbicular lobes, entire to shallowly or deeply emarginate. *Stamens* 5, with a very short filament and an oblong anther, inserted between the lower and upper third of the corolla-tube in long-styled (pin-eyed) corollas, in the

Map 1 Showing distribution of the genus *Dionysia*.

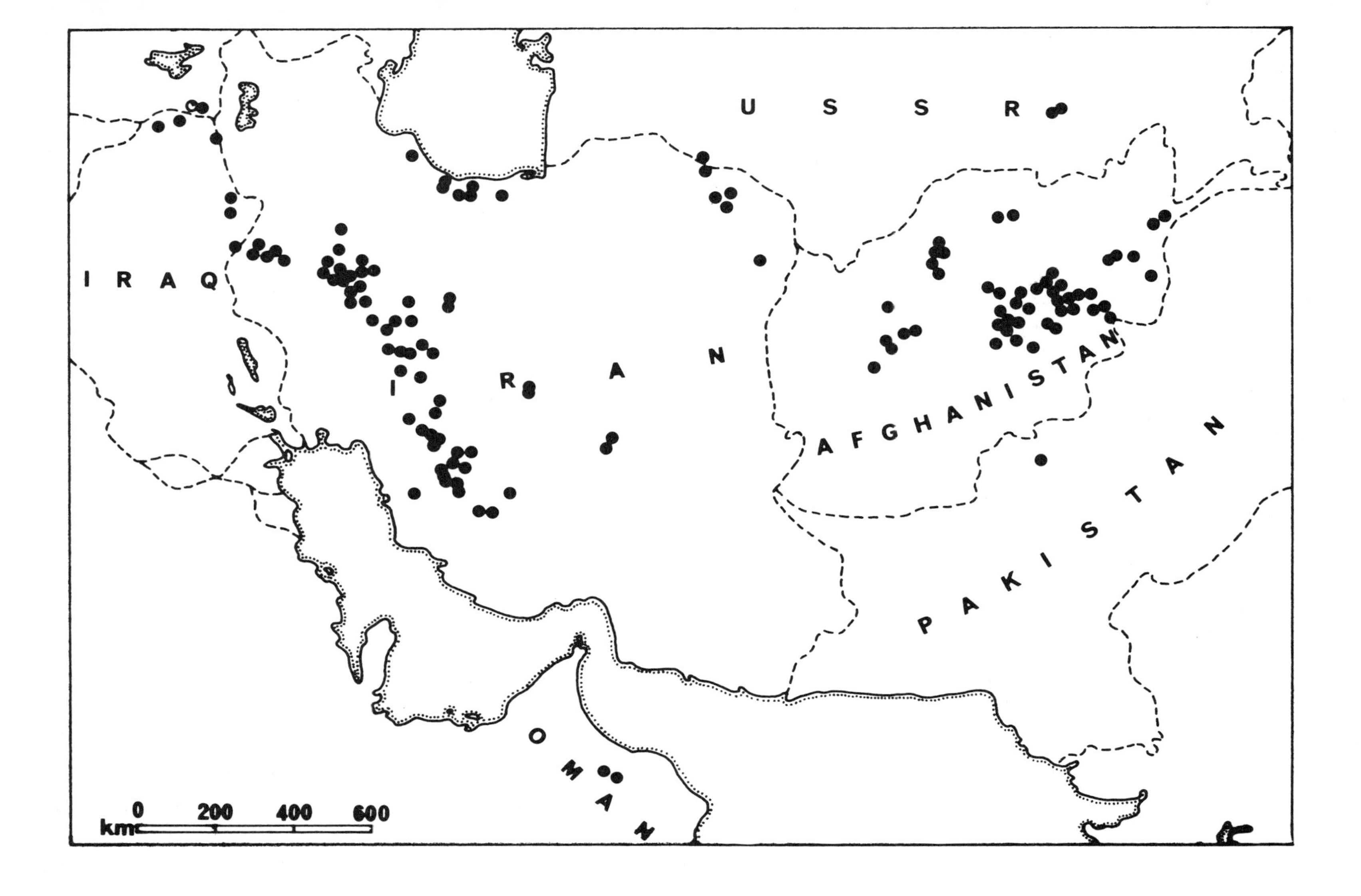
U S S R
I R A Q
I R A N
A F G H A N I S T A N
P A K I S T A N
O M A N
0
200
400
600
km

throat in short-styled flowers, rarely all inserted in the throat; pollen heteromorphic, ± depressed globose with 4–10 furrows, those from short-styled flowers larger and with more furrows than those from long-styled flowers. *Style* with a capitate stigma, reaching between one third and two thirds the way up the corolla-tube in short-styled (thrum-eyed) corollas or reaching the throat to exserted in long-styled (pin-eyed) corollas. *Fruit capsule* subglobose, splitting into 5 valves when mature, generally about half as long as the persistent calyx. *Seeds* few to numerous, larger when only a few to each capsule.

DISTRIBUTION. 41 species distributed from the extreme SE of Turkey and NW. Iraq to Iran, N Oman, S USSR (Kopet Dagh and Pamir Mts.), Afghanistan and Pakistan (Baluchistan).

HABITAT. In mountain regions, most frequently growing on cliffs of limestone, conglomerate or sandstone, occasionally granite, in shaded or partially shaded places, rarely in full sun, sometimes on sloping rock slabs or in cave entrances.

25 *Dionysia lamingtonii*. Photo. E. G. Watson.

26 *Dionysia lamingtonii*. Photo. E. G. Watson.

27 Kuh-i-Bamu, the type locality of *D. michauxii*, S. Iran. Photo. J. C. Archibald.

28 *Dionysia michauxii* growing on limestone cliffs, Kuh-i-Bamu, S. Iran. Photo. J. C. Archibald.

Sectional Classification

The genus can be divided into 3 main sections and a number of subsections, as follows:

Section Anacamptophyllum Melchior
Young leaves with revolute vernation. Mature leaves (of flowering shoots) 15 mm. long or more, with a flat, toothed margin, when smaller with a flat, lobed or distinctly revolute margin. Farina usually present and of the woolly type.

Subsection Scaposae Wendelbo. Leaves large or small, with flat margins at maturity (except *D. teucrioides*), with woolly farina beneath. Inflorescence scapose (almost obsolete in *D. saponacea*), with 1-several superposed whorls of flowers. Corolla yellow with entire lobes.

Subsection Revolutae Wendelbo. Leaves small, revolute at maturity, usually with woolly farina beneath. Inflorescence reduced to 1(-3) flowers; scape ± obsolete. Corolla yellow, pink or violet, with entire or emarginate lobes.

Section Dionysia Fenzl
Young leaves flat or with very slightly involute vernation. Mature leaves (of flowering shoots) small, mostly less than 10 mm long, with a flat margin, entire in most species (except in subsections *Caespitosae* and *Tapetodes*). Farina rare (present in *D. lurorum*, *D. denticulata* and *D. tapetodes*).

Subsection Caespitosae Wendelbo. Leaves spreading, not closely imbricating, generally with a toothed margin, with pinnate veins. Inflorescence with or without a scape. Corolla yellow, glandular-pubescent, with entire lobes (except in *D. termeana*); anthers inserted in the upper quarter of the corolla-tube in the pin-eyed flowers in which the style is clearly exserted.

Subsection Bryomorphae Wendelbo. Leaves closely imbricating, entire, with pinnate or occasionally flabellate veins. Flowers solitary, sessile. Corolla yellow, pink or violet, usually glandular or pubescent, occasionally glabrous, with emarginate, more rarely entire lobes; anthers inserted near middle of corolla-tube in pin-eyed flowers and the style never exserted.

Subsection Heterotrichae Wendelbo. Leaves not closely imbricating, with pinnate veins. Flowers solitary, sessile. Corolla violet, with emarginate lobes, glabrous; anthers inserted in middle of corolla-tube in pin-eyed flowers and style never exserted.

Section Dionysiastrum Smoljan.
Young leaves flat or with very slightly involute venation. Mature leaves, small, mostly less than 10 mm long, with a flat margin, toothed or entire, mostly with raised veins. Farina generally absent (present in *D. hedgei*, *D. microphylla* and *D. involucrata* and of the powdery type). Corolla pink or violet.

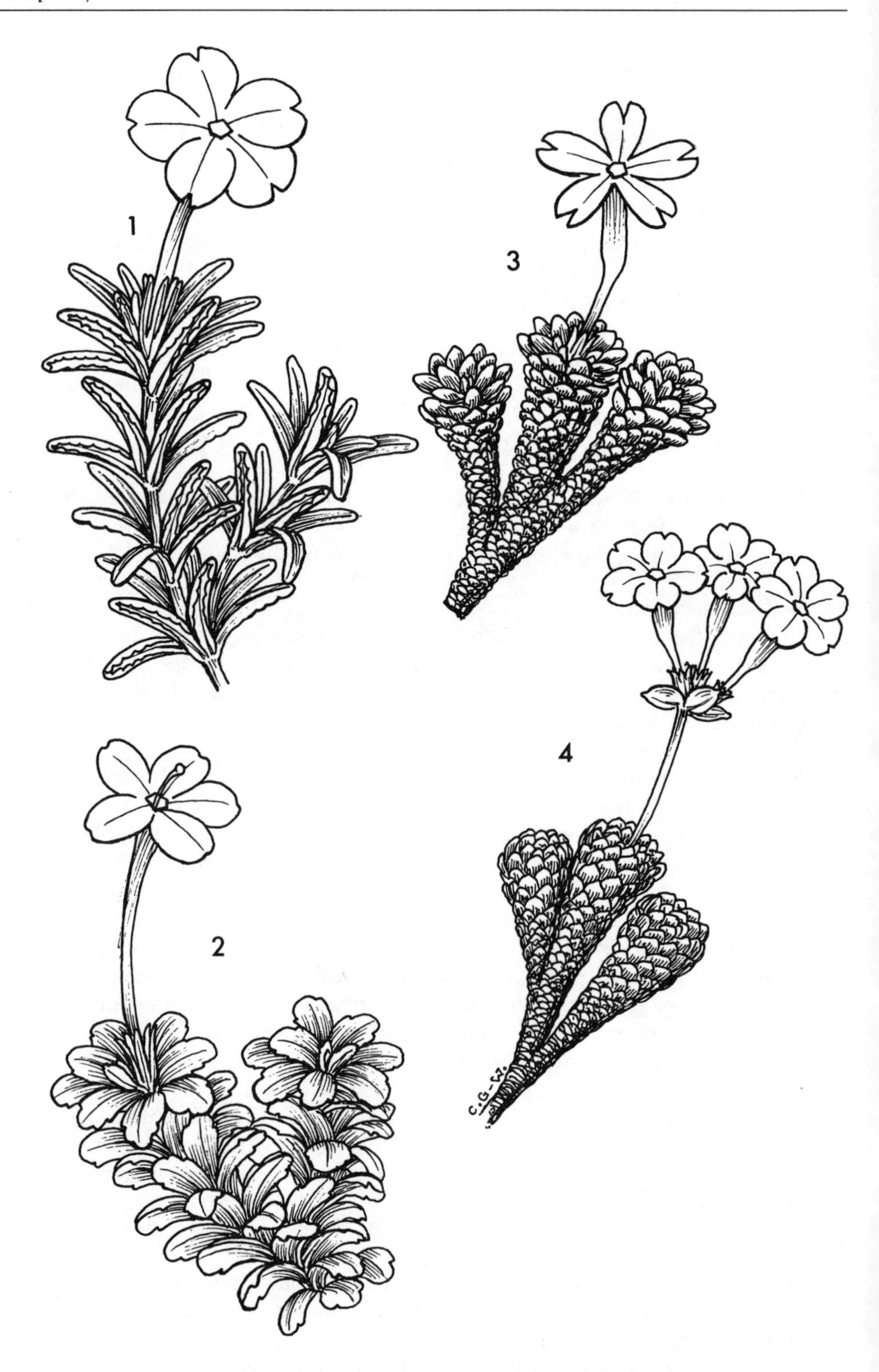

Diagram illustrating mode of growth and leaf arrangements in 4 species of *Dionysia*. 1 *D. aretioides*; 2 *D. diapensiifolia*; 3 *D. curviflora*; 4 *D. microphylla*. All × 3.

Subsection Involucratae Wendelbo. Leaves spreading, not closely imbricating, with distinct raised veins, especially on the upper surface. Flowers solitary, sessile, or in scapose inflorescences. Corolla with emarginate or entire lobes.

Subsection Afghanicae Grey-Wilson. Leaves closely imbricating, with obscure veins, not raised above. Flowers solitary, sessile. Corolla with emarginate lobes. (See Appendix, p. 168)

Subsection Microphyllae Wendelbo. Leaves closely imbricating, scale-like, with obscure veins. Inflorescence scapose. Corolla with slightly emarginate lobes.

Section ANACAMPTOPHYLLUM

Subsection SCAPOSAE

1. D. mira
2. D. bornmuelleri
3. D. teucrioides
4. D. balsamea
5. D. paradoxa
6. D. lacei
7. D. saponacea
8. D. hissarica

Subsection REVOLUTAE

9. D. aretioides
10. D. leucotricha
11. D. revoluta
12. D. archibaldii
13. D. rhaptodes
14. D. oreodoxa
15. D. esfandiarii

Section DIONYSIA

Subsection CAESPITOSAE

16. D. lurorum
17. D. caespitosa
18. D. gaubae
19. D. diapensiifolia
20. D. odora
21. D. termeana

Subsection BRYOMORPHAE

22. D. sawyeri
23. D. haussknechtii
24. D. lamingtonii
25. D. michauxii
26. D. curviflora
27. D. janthina
28. D. iranshahrii
29. D. bryoides
30. D. zagrica
31. D. sarvestanica
32. D. denticulata
33. D. tapetodes
34. D. kossinskyi

Subsection HETEROTRICHAE

35. D. lindbergii

Section DIONYSIASTRUM

Subsection INVOLUCRATAE

36. D. involucrata
37. D. hedgei
38. D. freitagii
39. D. viscidula

Subsection AFGHANICAE

40. D. afghanica

Subsection MICROPHYLLAE

41. D. microphylla

Key to species of Dionysia

1. Scape distinct, at least 10 mm long and generally much longer; inflorescence consisting of an umbel, or superposed verticils of flowers — 2
   Scape generally obsolete, if present less than 10 mm long; inflorescence generally reduced to a solitary flower (except in *D. caespitosa*, *D. lacei* and *D. microphylla* — 11

2. Corolla pink or violet — 3
   Corolla yellow — 5

3. Leaves small, not more than 1.5 mm long, without prominent raised (flabellate) venation; corolla-limb 8–9 mm diameter — 41. **D. microphylla***
   Leaves larger, 4–9 mm long, with prominent raised venation on the upper surface; corolla-limb 9–14 mm diameter — 4

4. Corolla-lobes emarginate; leaves generally with several small blunt teeth at apex — 36. **D. involucrata***
   Corolla-lobes entire, very rarely slightly emarginate; leaves entire — 37. **D. hedgei**

5. Plants forming ± dense cushions, the leaves of flowering shoots not more than 6 mm long, obscurely toothed to entire; corolla-lobes suborbicular. — 17. **D. caespitosa**
   Plants forming lax tufts, the leaves all more than 8 mm long, markedly toothed; corolla lobes narrow-ovate to elliptical — 6

6. Leaves of flowering shoots in distinct whorls — 8. **D. hissarica**
   Leaves of flowering shoots not in distinct whorls — 7

7. Scape up to 3 cm long, bearing a single umbel of flowers; leaves of flowering shoots to 12 mm long, margin distinctly revolute — 3. **D. teucrioides***
   Scape 3 cm long or more, generally bearing 2 or more whorls of flowers one above the other; leaves of flowering shoots 2 cm long or more, the margin not or scarcely revolute — 8

8. Leaves not pubescent, though covered in minute stipitate glands — 5. **D. paradoxa***
   Leaves with both long articulated hairs and minute stipitate glands — 9

9. Corolla-limb 12–18 mm diameter — 4. **D. balsamea***
   Corolla-limb 6–8 mm diameter — 10

10. Leaves oblong to oblanceolate; corolla-lobes never reflexed — 1. **D. mira***
    Leaves obovate to spathulate; corolla-lobes strongly reflexed, at least at first — 2. **D. bornmuelleri***

11. Leaves of mature flowering shoots with distinctly revolute margins — 12
    Leaves of mature flowering shoots with flat margins — 18

12. Corolla pink or violet — 13
    Corrolla yellow — 14

13. Leaves of flowering shoots c 4 mm long, the margin revolute only in the upper half, with long spreading articulated hairs towards the top — 15. **D. esfandiarii**
    Leaves of flowering shoots 4–7.5 mm long, the margin revolute for most of the leaf-length, covered in minute stipitate glands, but without articulated hairs — 12. **D. archibaldii***

14. Corolla-lobes entire — 15
    Corolla-lobes deeply emarginate — 16

15. Leaves 2.5–3.5 mm long, the margin entire, sparsely to densely covered with articulated hairs — 13. **D. rhaptodes**
    Leaves 4–5.5 mm long, the margin with 3–4 pairs of small teeth towards the leaf-apex, with stipitate glands, but without articulated hairs — 14. **D. oreodoxa**

16. Leaf-margins with about 6–8 pairs of small obtuse teeth — 11. **D. revoluta***
    Leaf-margins with about 3–4 pairs of coarser obtuse teeth — 17

17. Calyx-lobes entire, densely pubescent to the apex — 9. **D. aretioides***
    Calyx-lobes dentate at the apex, hyaline and almost glabrous towards the apex — 10. **D. leucotricha**

18. Leaves distinctly revolute when young, the margin coarsely toothed — 19
    Leaves flat or very slightly involute when young, the margin entire or somewhat toothed — 20

19. Inflorescence reduced to a solitary flower; leaves with curved articulated hairs — 7. **D. saponacea**
    Inflorescence of 4–8 flowers, practically scapeless — 6. **D. lacei**

20. Scape present but very short, 2–10 mm long; flowers 3 or more usually — 21
    Scape obsolete; flowers solitary, occasionally 2 together — 23

21. Corolla violet; leaves not more than 1.5 mm long — 41. **D. microphylla***
    Corolla yellow; leaves 2.5–6 mm long — 22

22. Plants with woolly farina; style of long-styled flowers not exserted — 16. **D. lurorum**
    Plants efarinose; style of long-styled flowers clearly exserted — 17. **D. caespitosa**

23. Corolla pink or violet — 24
    Corolla yellow — 33

24. Leaves with distinct raised (flabellate) venation on the upper surface — 25
    Leaves with rather indistinct venation — 27

25. Corolla-limb 5–6 mm diameter; leaves 2–2.5 mm long — 34. **D. kossinskyi**
    Corolla-limb 7–12 mm diameter; leaves 5–8 mm long — 26

26. Leaves elliptic-spathulate to oblong-oblanceolate; calyx divided for three-quarters of its length; corolla-tube 8–10 mm long 39. **D. viscidula***
    Leaves broadly elliptic to subrhombic; calyx divided for a half its length; corolla-tube 13–16 mm long 38. **D. freitagii***

27. Leaves and calyces finely stipitate glandular, without articulated hairs 28
    Leaves and calyces with long articulated hairs, sometimes also with stipitate glands 29

28. Leaf-apex truncate or slightly emarginate; leaves 1 mm wide or more; calyx divided for two thirds of its length 40. **D. afghanica***
    Leaf-apex obtuse to subobtuse; leaves up to 1 mm wide; calyx divided to the base 29. **D. bryoides***

29. Corolla-lobes entire 28. **D. iranshahrii**
    Corolla-lobes emarginate 30

30. Leaves linear-spathulate, not closely imbricate, with a mixture of long articulate hairs and stipitate glands; plants forming soft grey cushions 36. **D. lindbergii**
    Leaves oblong to elliptical, closely imbricate, without stipitate glands; plants forming rather dense hard green or silver-grey cushions 31

31. Leaves densely pubescent beneath; calyx pubescent outside 27. **D. janthina***
    Leaves glabrous beneath except for the margins; calyx glabrous outside except for the margins of the lobes 32

32. Leaves c.5 mm long; bract solitary; corolla-limb 8–9 mm diameter 22. **D.sawyeri**
    Leaves 2–3 mm long; bracts two; corolla-limb to 6 mm diameter 26. **D. curviflora***

33. Marcescent leaves spreading, not closely imbricate 34
    Marcescent leaves closely imbricate 35

34. Leaves with 2 pairs of regular teeth at the apex, 4–6 mm long 18. **D. gaubae**
    Leaves irregularly and often deeply toothed, 5–10 mm long 19. **D. diapensiifolia***

35. Leaves and calyces generally with capitate glands, but without long articulated hairs 36
    Leaves and calyces with long articulated hairs, at least in part 39

36. Leaves 6–11 mm long, often in distinct whorls along the stems; corolla-limb 10–12 mm diameter 21. **D. termeana**
    Leaves 2–8 mm long, not in distinct whorls along the stems; corolla-limb 5–8 mm diameter 37

* species in cultivation

37. Corolla minutely glandular outside; calyx split for three-quarters or more of its length — 30. **D. zagrica**
Corolla glabrous outside; calyx split for a half to two-thirds of its length — 38

38. Cushions relatively lax; leaves mostly 4–8 mm long, often pronouncedly denticulate; corolla-lobes emarginate — 32. **D. denticulata***
Cushions relatively dense; leaves mostly 2–4 mm long, generally entire; corolla-lobes entire, rarely somewhat emarginate — 33. **D. tapetodes***

39. Leaves practically glabrous above, except for the margin; calyx divided for about half its length — 24. **D. lamingtonii***
Leaves pubescent or stipitate-glandular above; calyx divided almost to the base — 40

40. Leaf fringed with articulated hairs, the upper and lower surface with stipitate glands only; cushions grey-green — 23. **D. haussknechtii**
Leaf covered all over with long articulated hairs; cushions silvery-grey — 25. **D. michauxii***

The species

1 *Dionysia mira*

Dionysia mira Wendelbo in Bot. Not. 112: 495 and in Årbok Univ. 1, Berg., Mat.–Nat. Ser. No. 3: 36 (1961). Type: Oman, Mt Akadar (Jebel Akdar), *Aucher-Eloy* 5236 (holotype P; isotypes BM, G, K).

Primula aucheri Jaub. & Spach, Illustr. Pl. Orient 1, 43: 97, t. 49 (1842), non *Dionysia aucheri* (Duby) Boiss. (1879).

DESCRIPTION. *A suffrutescent perennial herb*, the plants forming large loose tufts up to 80 cm across; branches becoming woody with age and covered with marcescent leaves and leaf-bases, eventually more or less bare. *Leaves* revolute when young, oblong to lanceolate or oblanceolate, 2.5–8.3 cm long (including the petiole), 0.5–2.5 cm wide, the margin incised dentate to bi-dentate, covered in articulated hairs and minute glands on both surfaces, with white woolly farina beneath, especially when young. *Inflorescence* solitary or 2–3 per leaf rosette, each composed of 3–7 superposed whorls of 5–7 flowers; flowers heterostylous. *Peduncle* 6–21 (–30) cm long, including the rachis, pubescent like the leaves. *Bracts* foliaceous, variable, the lower lanceolate, serrate to incised-dentate, as long as or exceeding the adjacent flower, the upper bracts lanceolate to sub-linear, ± entire, all pubscent as the leaves. *Calyx* narrowly campanulate, 7–15 mm long, divided half-way into linear-triangular acute to long acuminate lobes, glandular, pubescent, sometimes with some woolly farina, especially near the base. *Corolla* yellow, glandular-pubescent outside; tube 15–18 mm long; limb 7–9 mm diameter, divided into oval to ovate, entire to slightly emarginate, lobes. *Anthers* c.2 mm long, inserted in the throat or about half-way along the corolla-tube. *Style* reaching the throat or half-way along the corolla-tube. *Capsule* with numerous seeds (Plate 5).

DISTRIBUTION. Oman; Jebel Akdar, Jebel Aswad and Jebel Sharm – not known elsewhere; altitude 1500–1900 m.

HABITAT. Sloping limestone rocks or cliffs, often of north or north-east aspect, generally inhabiting rather open sites, or alternatively growing in partially wooded gullies amongst boulders close to streams. Flowering from January to April.

Dionysia mira was first discovered in 1838 by Aucher-Eloy. Since that date this interesting species has been seen and collected only a few times. Indeed the second collection was made only 30 years ago, in 1959, by Smiley and Deacock. In the 1970s several expeditions explored the mountains of northern Oman. Most notable of these were the collections of Alan Radcliffe-Smith from the Royal Botanic Gardens, Kew. From these collections, numbered ARS 4037, came the plants now in cultivation.

Dionysia mira is the most primula-like of any *Dionysia* species with its lax rosettes of quite large leaves and tiered whorls of yellow flowers, borne on long

Dionysia mira, × 1. 1–2 calyces, × 3; 3 pin-eyed corolla, × 3; 4 thrum-eyed corolla, × 3.

scapes. It is also the only species to be found south of the Persian Gulf. Wendelbo placed this species in a subsection of its own (subsection *Mirae*), primarily on account of its bullate leaf-surface and powdery, not woolly, farina. These observations were made from rather scanty dried material. I have examined live material of *D. mira* and its closest ally, *D. bornmuelleri*, and there is little difference in the character of the leaf-surface. The farina, although less thread-like in *D. mira* is certainly of the woolly type, not the powdery Primula-type. I have no hesitation therefore in merging the two subsections *Mirae* and *Scaposae*, choosing the latter name as the more appropriate. Both subsections (of section *Anacamptophyllum* Melchior) were described by Wendelbo in 1961 in his monograph of the genus.

The dried collection of P.N. Munton at Kew (K), gathered in March 1978, differs from all the other material in the long, narrow calyx, 14–15 mm long (as opposed to 7–12 mm) with narrow, long-acuminate lobes. This plant was collected in a new locality, Jebel Aswad, and may perhaps be worthy of varietal recognition. However, at present not enough is known about the range of variability of the species.

Dionysia mira appears to have a rather limited distribution in the mountains of Oman, being found only in several refuges on Mt. Akadar (Jebel Akdar), where the type specimen came from, and Jebel Sharm. On these mountains it inhabits rather open sites, north and north-east facing limestone cliffs or sloping limestone rocks above 1500 m altitude.

A colour plate of *D. mira* occurs in Curtis's *Botanical Magazine*, t. 881 (1983), from a plant cultivated at the Royal Botanic Gardens, Kew.

This species was at first placed in the genus *Primula* by Jaub. & Spach who called it *P. aucheri*, in 1842. This is not to be confused with *Dionysia aucheri* (Duby) Boiss. a synonym of *D. odora* Fenzl (see p. 107). As a result when Wendelbo (1961) came to write his monograph of the genus he had to choose a new epiphet – *D. mira* Wend.

Dionysia mira, like its close cousins *D. bornmuelleri*, *D. teucrioides*, *D. paradoxa* and *D. balsamea*, all look superficially very similar to members of *Primula* section *Sphondylia* (syn. *Floribundae*), but they can be separated at once by the presence of revolute, rather than involute leaves, and the subshrubby, not herbaceous, habit; the leaf-characters have to be examined as the young leaves unfurl, when the revolution of the margins can be clearly observed.

Dionysia mira has proved to be an easy species in cultivation. Although the flowers are not particularly large they are produced plentifully, in succession, and make a pretty enough display. Its culture has been likened to that of certain Primulas, e.g. *P. floribunda* and *P. forrestii*; plants thriving in a gritty well-drained compost, and should be kept relatively dry in the late autumn and winter, increasing the supply of water as growth starts in the spring. Care should be taken not to water the foliage in winter and persistent dead leaves are best removed in the autumn, or indeed as they occur, as they can be the site of fungal infection. Plants can be readily propagated from cuttings (stem or leaf), by division of the parent plant or by seed; both pin- and thrum-eyed plants are in cultivation. Cuttings give a success rate of 50–100%. Plants (from seed) grow rapidly, reaching 20 cm across in only 2 years.

Cultivated plants come into flower in late January (sometimes as early as late

December) and flower through to May, but occasional flowers occur throughout the summer.

Plants planted out in borders of the alpine house at Kew have thrived for several years but need replacing regularly otherwise they eventually become too lax and less floriferous.

Dionysia mira has been successfully crossed with *D. aretioides*, see p.167.

2 *Dionysia bornmuelleri*

Dionysia bornmuelleri (Pax) Clay, The Present-day Rock Garden: 194 (1937).

Primula bornmuelleri Pax, Schles. Gesellsch. vaterl. Cul. II Abt. Naturw. b. Zool. – bot. Sekt. Jahresb. 87: 21 (1909). Type: Iran, Kermanshah, Noah Kuh near Kerind, *Strauss* 601 (holotype B; isotypes E, K, W).

DESCRIPTION. *Plants* forming large lax tufts to 50 cm across; branches becoming bare and woody with age, covered above with marcescent leaves and leaf-remains. *Leaves* forming a loose rosette, oblong to obovate or ± spathulate, revolute at first, 1–7.4 cm long including the petiole, 0.5–2.5 cm wide, the margin bidentate or bicrenate, venation reticulate, especially prominent beneath, covered on both surfaces with articulated hairs, generally with rather dense whitish farina beneath. *Inflorescence* solitary or 2 per leaf-rosette, composed of 2–4 superposed whorls, rarely a simple umbel, each with 3–5 flowers; flowers heterostylous. *Peduncle* 2.8–10 cm long, including the rachis, pubescent. *Bracts* foliaceous, ovate-lanceolate to oblong or lanceolate, 1–2.5 cm long, coarsely toothed to ± entire, pubescent as the leaves. *Calyx* tubular-campanulate, 5–8 mm long, divided for three-quarters of its length into linear-lanceolate, acute, entire lobes, pubescent. *Corolla* yellow, often with a brown or orange spot at the base of each lobe, glandular-pubescent outside; tube 17–29 mm long; limb 5–8 mm diameter, divided into ovate, entire, lobes that are reflexed at first, later patent and flat. *Anthers* c. 2 mm long, inserted in the middle of the corolla-tube or just below the throat. *Style* reaching the middle of the corolla-tube or just below the throat. *Capsule* containing numerous seeds.

DISTRIBUTION. SE Turkey (Mardin-Cúdi Dağ); NE Iraq (Mosul District – near Sharanish, Erbil District – Rowanduz Gorge; W Iran (Kermanshah – near Kerind); altitude 500–1500 m.

HABITAT. Shady limestone rocks, especially cliffs, sometimes on tufa, especially where there is a water drip or seepage. Flowering March to May.

Dionysia bornmuelleri was first discovered by Th. Strauss in W Iran in 1909. Strauss recognised his find to be a new species of the genus *Dionysia* and he proposed that it be named after and in honour of J. Bornmüller who had named and worked on many of Strauss's collections from Iran. However, when Pax described the species in 1909 he placed it not in *Dionysia*, but instead described it as a new species of *Primula*, *P. bornmuelleri*. It was not until 1937 that Clay in *The Present-day Rock Garden* formally recognised this plant as a member of *Dionysia*. From anatomical, pollen and seed morphology studies this plant clearly

Dionysia bornmuelleri, × 1. 1 calyx, × 3; 2 thrum-eyed corolla, × 2; 3 pin-eyed corolla, × 2. *D. teucrioides*. 4 vegetative leaf, × 2; 5 leaf of flowering shoot, × 2; 6 calyx, × 3; 7 corolla, × 2.

belongs in *Dionysia*, not *Primula*.

Dionysia bornmuelleri is rather a rare species in cultivation. This is not so much because it is particularly difficult in cultivation but because it is a rather lax plant with small slender-tubed flowers which are less striking than most of its cousins. However, the plant has a certain elegance and deserves a place in any collection of Dionysias. The flowers, though long-tubed, have a very small limb. When they first open the corolla-lobes are strongly deflexed, but shortly they come forward to form a more or less flat limb.

Plants grow equally well in either gritty compost or in tufa. Certainly in the latter they make rather more shapely plants.

Cuttings taken in June and July root fairly readily and one can expect a 'take' of 60–80 per cent. Plants may reach a diameter of about 30 cm in 6 or 7 years.

Plants in cultivation originated from seed collected in 1965 in Iraq by Professor Rechinger of Vienna and sent to Per Wendelbo at Göteborg (under *Rechinger* 11485) and all the plants in cultivation today come from this one source.

Per Wendelbo wrote in 1969 – '*D. bornmuelleri* has flowered with us (at Göteborg). It has very small flowers, but the tube is long. The colour is orange-yellow with orange markings near the base of the lobes' and '... we manage to keep *D. bornmuelleri* and *D. parodoxa* in the alpine house with great difficulty, although the former has now been kept for nearly 10 years'.

Botrytis can sometimes infect plants. Removal of dead basal leaves can help to reduce possible attacks.

Dionysia bornmuelleri has been hybridised with *D. aretioides* in cultivation – see p. 167.

3 *Dionysia teucrioides*

Dionysia teucrioides Davis & Wendelbo in Årbok Univ. I. Berg. Mat.-Nat. Ser. 3, I: 39 and 76 (1961). Type: Turkey, Hakkari, Cilo Dağ, Aug. 1954, *Davis & Poulunin* D. 23884 (holotype K; isotype E).

DESCRIPTION. *Dwarf suffruticose perennial* forming rather dense tufts to 20 cm across; branches becoming bare and woody below, above covered with reflexed marcescent leaves and leaf-bases. *Leaves* in lax rosettes, revolute, occasionally becoming flat, oblong to oblanceolate, 8–12 mm long including the petiole, 2.5–4 mm wide, the revolute margin with 4–6 obtuse simple teeth on each side, densely covered on both surfaces by articulated hairs and short capitate glands and beneath with rather dense-whitish woolly farina; leaves of vegetative shoots larger, to 27 mm long and 11 mm wide, flat-margined, with up to 8 teeth on each side. *Inflorescence* a 1–3 flowered umbel, occasionally with a superposed whorl of 1 flower or 1 bract; flowers apparently homostylous. *Peduncle* 1–3 cm long, glandular-pubescent. *Bracts* foliaceous, oblong, 5–9 mm long, serrate in the upper half. *Calyx* tubular-campanulate, 6–7 mm long, divided for three-quarters of its length into lanceolate, acute, entire lobes, glandular-pubescent. *Corolla* yellow, somewhat glandular-pubescent outside; tube 16–18 mm long; limb 5–7 mm diameter, divided into ovate, entire lobes. *Anthers* c. 1.5 mm long; inserted in the throat of the corolla. *Style* reaching three-quarters the way along the corolla-tube initially, but extending eventually beyond the anthers and exserted. *Capsule* containing numerous seeds. (Plate 6).

DISTRIBUTION. SE Turkey, Hakkari Province, Cilo Dağ – not known elsewhere; altitude 1900 m.

HABITAT. Growing beneath overhanging rocks on limestone cliffs, in dry shaded places. Flowering April-May.

Dionysia teucrioides is endemic to the extreme south-eastern corner of Turkey. Indeed it is the only species of *Dionysia* endemic to Turkey.

Dionysia teucrioides was first discovered in 1954 by Peter Davis and Oleg Polunin in Hakkari Province, on Cilo Dağ, a mountain close to the Iran/Iraq border. This species was found growing in shady, rock underhangs. Peter Davis writes – '*D. teucrioides* grows in SE Anatolia (Diz Gorge on Cilo Dağ and the cliffs of Cudi Dağ, above Gizze), always on overhanging limestone cliffs in partial

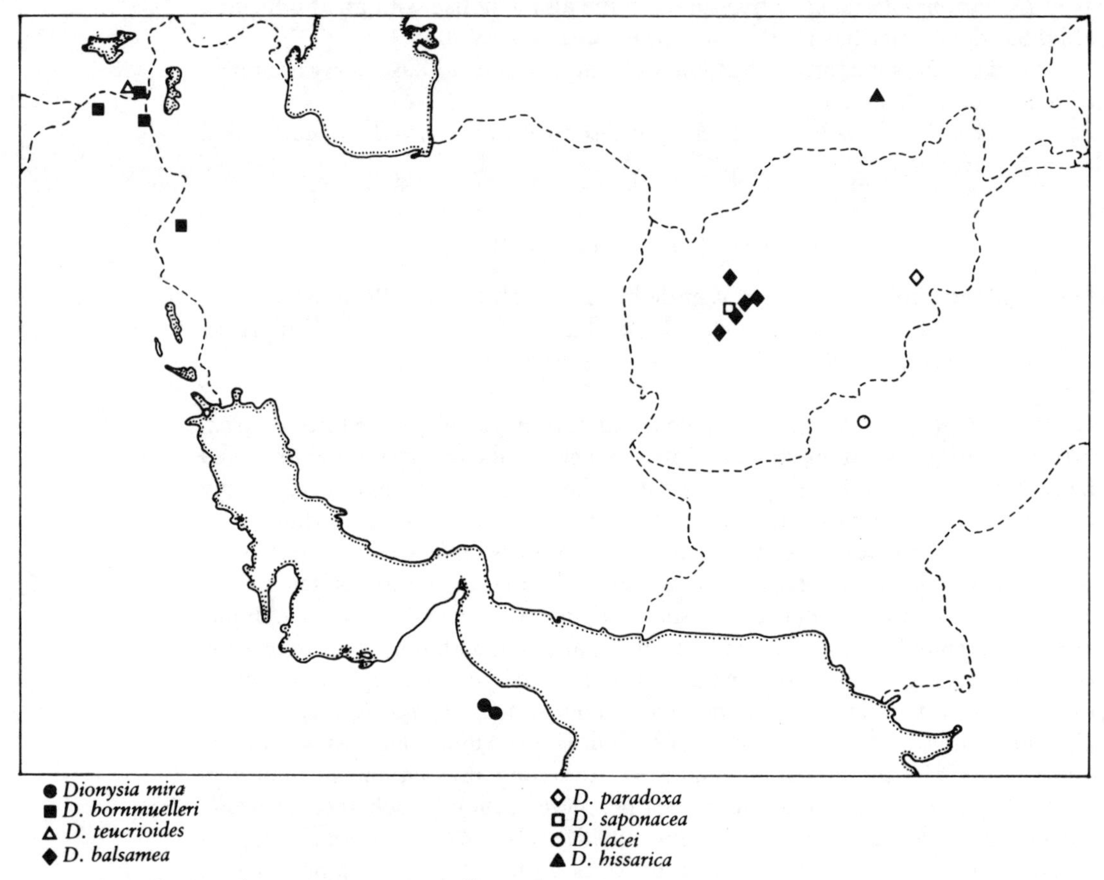

Map 2 Distribution of section *Ancamptophyllum* subsect. *Scaposae*.

shade, and close to *Primula davisii*'. It is rather extraordinary that the *Primula* and *Dionysia* should grow side by side, for the yellow-flowered *Primula davisii* is placed in *Primula* section *Sphondylia*, to which *Dionysia* is related, see p. 26.

Dionysia teucrioides finds its closest ally in *D. bornmuelleri*, indeed, the latter has a geographical distribution just to the south of *D. teucrioides*. There is little doubt that *D. teucrioides* has been derived from *D. bornmuelleri* at some time. It can be distinguished by its smaller leaves which more or less maintain revolute margins to maturity, by the shorter scapes bearing fewer (1–3) somewhat smaller flowers. In many respects *D. teucrioides* comes midway between *D. bornmuelleri* and species of section *Revolutae*, especially *D. aretioides* and *D. leucotricha*, although with its scapes and foliaceous bracts it clearly 'fits' into section Scaposae.

Dionysia teucrioides occurs at considerably higher altitudes than *D. bornmuelleri* – 1900 m as against 500–1000 m.

This interesting species was introduced into cultivation in 1967 from seed collected by John Watson (under ACW 3619) and the first plant to flower did so in the early spring of 1969. However, like the related *D. bornmuelleri* it proved rather difficult in cultivation. Jack Elliott wrote in 1969: 'This germinated well from ACW seed, but I had another disaster. Six seedlings in one plant pot died after repotting, presumably from damping off. I had another pot with two seedlings, and these I was able to pot on and they have done reasonably well ... However, Per Wendelbo wrote at the same time: '*D. teucrioides* is growing well with us and we have quite a few plants'. The species was initially distributed under the incorrect name of *D. teucrifolia*.

Once nurtured beyond the seedling stage plants grow reasonably fast, attaining 10 cm across in 4 years, 17 cm in 6 years. Plants form loose low tufts supporting up to 30 leaf-rosettes. The flowers, which are produced during February and March, are rather small and sparse and the species is perhaps of more botanical than horticultural interest. However, a well-grown plant is attractive enough.

Plants are not easy from cuttings; they have a low success rate and those that root are rather slow to 'get away'. Seed has been produced naturally in cultivation, apparently from a single clone – it may take up to 3 years to germinate.

Botrytis and aphids, the two chief dangers to any *Dionysia*, can prove bothersome and need to be watched out for carefully, especially in autumn, winter and early spring.

Dionysia teucrioides has been hybridised with *D. aretioides*, see p. 167.

4 *Dionysia balsamea*

Dionysia balsamea Wendelbo & Rech. fil. in Årbok Univ. I. Berg. Mat.-Nat. Ser. 19, IV: 7, figs. 2, 3 e-f (1964). Type: Afghanistan, Ghorat Province, Kuh-Tscheling-Safed-Daraq (Pirestan), July–August 1962, *Rechinger* 19094 (holotype W; isotype BG).

Description. *A laxly tufted suffrutescent* perennial rather like *D. paradoxa*, forming tufts up to 50 cm across; branches becoming woody and leafless below,

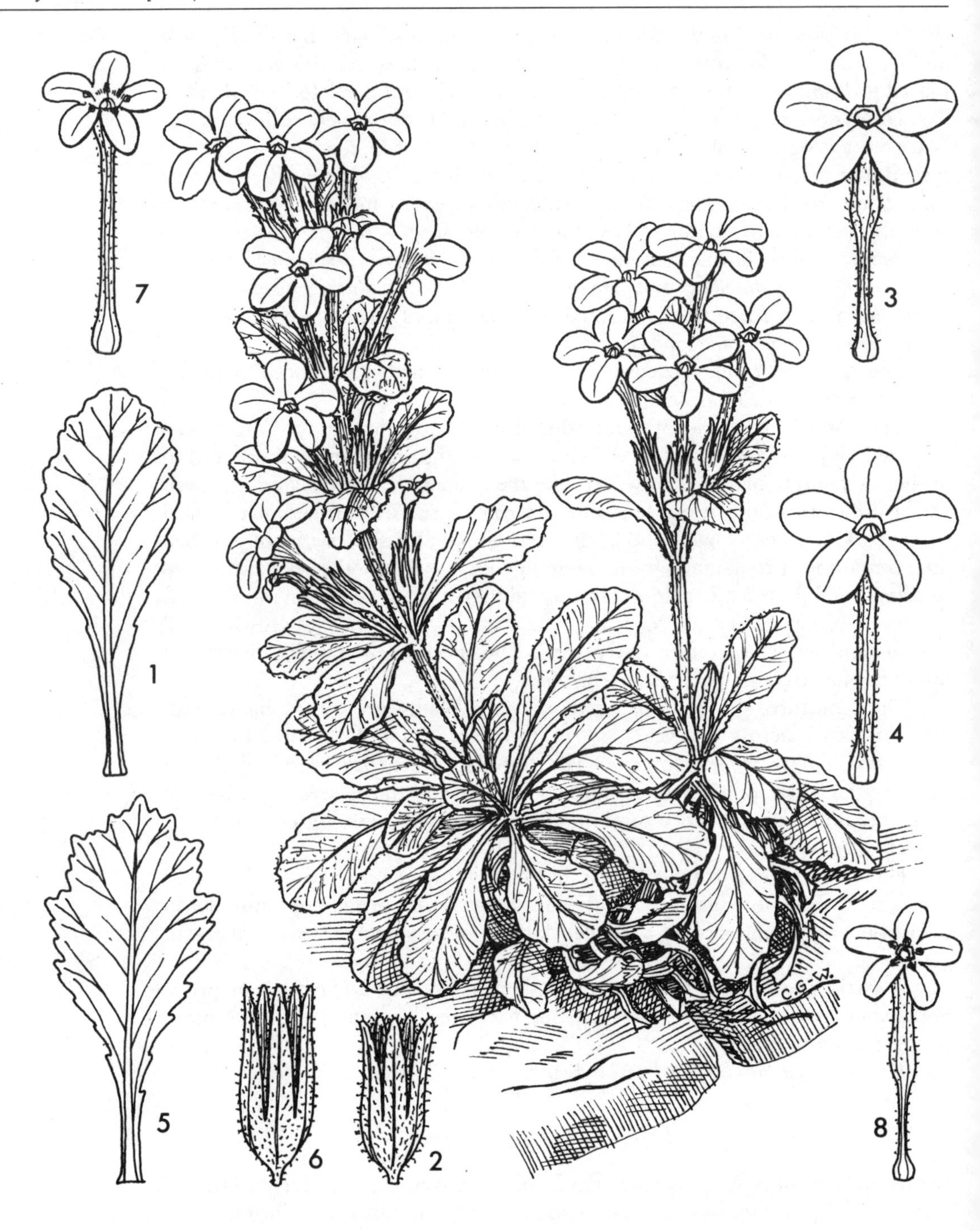

Dionysia balsamea, × 1. 1 leaf, × 1½; 2 calyx, × 3; 3 pin-eyed corolla, × 2; 4 thrum-eyed corolla, × 2. *D. paradoxa*. 5 leaf, × 1½; 6 calyx, × 3; 7 thrum-eyed corolla, × 2; 8 pin-eyed corolla, × 2.

29 *Dionysia curviflora* growing on limestone cliffs, Schir Kuh, WC Iran. Photo. J. C. Archibald.

30 *Dionysia curviflora*. Photo. S. Taylor.

31 *Dionysia janthina*, growing on cliffs S. of Yazd, WC Iran. Photo. J. C. Archibald.

32 *Dionysia janthina*, growing on cliffs S. of Yazd, WC Iran. Photo. J. C. Archibald.

with marcescent leaves and leaf-remains above. *Leaves* forming a loose rosette, oval to obovate or obovate-spathulate, 2–5.2 cm long including the petiole, 0.7–2.2 cm wide, the apex obtuse, the margin crenate to crenate-dentate or bi-crenate, covered on both surfaces with glandular-hairs, sometimes also with articulated hairs, with whitish woolly farina beneath. *Inflorescence* solitary or 2–3 per leaf-rosette, each composed of 1–4 superposed whorls each of 3–5 flowers; flowers heterostylous. *Peduncle* 3.5–10.2 cm long, including the rachis, glandular-pubescent. *Bracts* foliaceous, oval to obovate, 1.7–2.8 cm long, rather coarsely crenate, glandular-pubescent like the leaves. *Calyx* tubular-campanulate, 7–9 mm long, divided for about two-thirds of its length into linear-lanceolate, slightly toothed lobes. *Corolla* pale to deep yellow, glandular outside; tube 15–25 mm long; limb 10–18 mm diameter, divided into broad obovate or oval entire lobes. *Anthers* c. 2 mm long, inserted in the throat or near the middle of the corolla-tube. *Style* reaching the throat or the middle of the corolla-tube. *Capsule* with 5–6 seeds (Plates 7–8).

DISTRIBUTION. WC Afghanistan, Ghorat, Farah and Herat provinces, scattered localities from Kuh-Tscheling-Safed-Daraq to Parjuman, SW Naourak, SW of Taiwara, Hari Rud valley near Chisht, NE of Gulestan and Djam, W of Chagcharan; altitude 1500–2800 m.

HABITAT. Limestone rock crevices, cliffs and cliff ledges, growing in shaded or partially shaded places. Flowering during February and March.

Dionysia balsamea was first collected by Professor K. H. Rechinger of Vienna in late July 1962 near Naourak in central west Afghanistan. Since that date it has been collected on a number of mountains in the same part of Afghanistan.

Dionysia balsamea has obvious affinities with *D. paradoxa*, which has a more limited distribution in the east of Afghanistan. Although the two species overlap in all dimensions, *D. balsamea* is a generally coarser plant, the leaves are proportionately broader and blunter with less sharp teeth and the same applies to the bracts. Furthermore, the leaves and bracts of *D. paradoxa* have very short capitate glands, 0.1 mm long, whereas those of *D. balsamea* are up to 1 mm long. In most instances the corollas of *D. balsamea* are larger.

Dionysia balsamea is one of the primula-like species, forming a large rather lax plant with leafy rosettes and tiers of bright yellow, quite large flowers, borne on erect scapes. When well grown this is an extremely handsome and floriferous species, well worth growing and probably the best of the primula-types.

Plants grow rather rapidly, reaching about 25 cm across, occasionally more, in about 5 years. Only the thrum-eyed form is in cultivation so that no seed has been produced. However, new plants are reasonably easily raised from cuttings struck in a gritty compost, or more successfully in fine pumice.

Like most of the more vigorous Dionysias, *D. balsamea* is a greedy plant requiring regular feeding. Plants, like *D. aretioides*, eventually get too large to handle and a new stock should then be raised. Potted plants can be grown in frames open to the elements during the summer months, covered but well ventilated during the autumn and winter.

Despite its relative ease in cultivation *D. balsamea* is still rare in collections, indeed at one time it almost died out in cultivation altogether. Plants in cultivation are all derived from the Grey-Wilson/Hewer 1973 Afghanistan expedition under the number GW/H 580. The seed was gathered from plants growing on cliffs within sight of the extraordinary thirteenth century Minaret of Djam, which was only discovered by the outside world in the 1960s.

5 *Dionysia paradoxa*

Dionysia paradoxa Wendelbo in Bot. Not. 112: 497 (1959). Type: Afghanistan, Kabul River, Sarobi, *Volk* 2411 (holotype BG).

Dionysia hissarica Wendelbo in Symbolae Afghanicae IV. Biol. Skr. Dan. Vid. Selsk. 10(3): 68 *non* Lipsky (1900).

DESCRIPTION. *A laxly tufted suffrutescent perennial herb*, the plants forming spreading mats up to 1 m across, though often less; branches becoming woody with age, bare below but with marcescent leaves and leaf-bases above. *Leaves* forming loose rosettes, revolute when young, obovate to elliptic or lanceolate, 2–4.8 cm long, 0.8–2.4 cm wide, the apex acute or sub-acute, the margin serrate to dentate or bidentate, rather unevenly so, densely covered on both surfaces with short glandular-hairs, with greyish woolly farina beneath. *Inflorescence* solitary or 2–3 per leaf-rosette, each composed of 2–4-superposed whorls of 2–4 flowers; flowers heterostylous. *Peduncle* 2–7 cm long, including the rachis, glandular-pubescent. *Bracts* foliaceous, lanceolate to oblong-lanceolate, 1.5–3 cm long, coarsely serrate or dentate, glandular-pubescent like the leaves. *Calyx* tubular-campanulate, 7–10 mm long, divided for about two-thirds of its length into linear-lanceolate, toothed lobes. *Corolla* yellow, often with pale orange or brown markings at the base of the lobes, glandular outside; tube 16–20 mm long; limb 10–12 mm diameter, divided into obovate or oval, rather narrow, entire lobes. *Anthers* c. 2 mm long, inserted in the middle of the corolla-tube or just below the throat. *Style* reaching the throat or the middle of the corolla-tube. *Capsule* with numerous seeds (Plate 9).

DISTRIBUTION. E. Afghanistan, Sarobi Gorge, especially near Sarobi – not known elsewhere; altitude 1000–1200 m.

HABITAT. Conglomerate cliffs, in shaded or partially shaded places, especially beneath overhangs, often where the rocks are moist. Flowering during February and March.

Dionysia paradoxa was first collected by Neubauer on the Afghanistan side of the Khyber Pass, in 1949. Since then it has been seen and collected on a number of separate occasions – most notably Paul Furse (1966), Hedge, Wendelbo & Ekberg (1969) and Grey-Wilson & Hewer (1971). Seed from the latter two collections (PW 7496 & GW/H 1419) produced a number of plants which are still in cultivation today.

Although generally dismissed by alpine growers as a botanical curiosity, *D. paradoxa* can look quite attractive when well grown and floriferous, deserving a small space in any collection. The species is one of the primula-types having the general appearance of *D. balsamea* (see p. 65) but laxer and less sturdy in habit with smaller flowers.

Indeed *D. paradoxa* finds its closest ally in *D. balsamea*. Both are endemic to Afghanistan, the former from the east, the latter from the centre of the country. *D. balsamea* can be distinguished readily by the larger, coarser leaves and blunt, not pointed teeth and apex, the covering of longer glandular hairs and in the larger flowers – the corolla-limbs of *D. balsamea* are almost twice as large!

Dionysia paradoxa occurs on shaded conglomerate cliffs in the wild, centred on the Sarobi Gorge between Kabul and the Khyber Pass. It often inhabits cave entrances and underhangs where it sometimes forms substantial colonies.

In cultivation this species requires treatment similar to *D. mira* (p. 56), but even stricter. Botrytis is more of a problem and cuttings more reluctant to root. Sometimes a large plant can be divided. This is best attempted in May and June. Jack Elliott wrote in 1968 – 'PF (Paul Furse) seed germinated well but early losses were horrible and I now have two large plants, not too healthy looking, one of which flowered (*AGS Bull.* 36: 208–211, 1968). If it flowers again I will try and arrange a public appearance, as I am sure that they will not live long and so far I have had little success with cuttings'. Today there are very few plants in cultivation. Plants in cultivation flower in February and March.

6 *Dionysia lacei*

Dionysia lacei (Hemsley & Watt) Clay, The Present-day Rock Garden (London): 195 (1937).

Primula lacei Hemsley & Watt in Journ. Linn. Soc. Bot. 28: 298, 325, t.41 (1891). Type: Pakistan, Baluchistan, Torkhan, *Lace* 3648 (holotype K, isotypes BM, E).

Description. *A tufted perennial* forming rather lax cushions; branches becoming bare and woody below, with marcescent leaves and leaf-bases above. *Leaves* forming lax rosettes, obovate to obovate-spathulate, 10–20 mm long, 4–7 mm wide) (those of vegetative shoots larger, to 35 mm long, 4–7 mm wide) the apex obtuse, the margin with 3–5 teeth on each side, covered on both surfaces with curled articulated hairs, with white or yellowish woolly farina beneath. *Inflorescence* reduced to a whorl or two closely set whorls, each with 3–5 flowers, the scape more or less obsolete; flowers heterostylous. *Peduncle* 1–3 mm long, obscure. *Bracts* linear-lanceolate to linear, 7–15 mm long, acute, pubescent. *Calyx* tubular-campanulate, divided for two-thirds of its length into linear-lanceolate entire lobes, pubescent. *Corolla* yellow, pubescent outside; tube 14–22 mm long; limb 12–18 mm diameter, divided into oblong to obovate, generally slightly emarginate, lobes. *Capsule* with ? seeds.

Distribution. Pakistan; Baluchistan, Torkhan, c. 120 km E of Quetta – not known elsewhere; altitude c. 1200 m.

Habitat. Limestone cliffs, growing in shady crevices. Flowering March–April.

In the protologue Hemsley and Watt (1891) remark – '*Primula lacei* is one of the most interesting, being the only *Primula* found hitherto in Baluchistan, and it

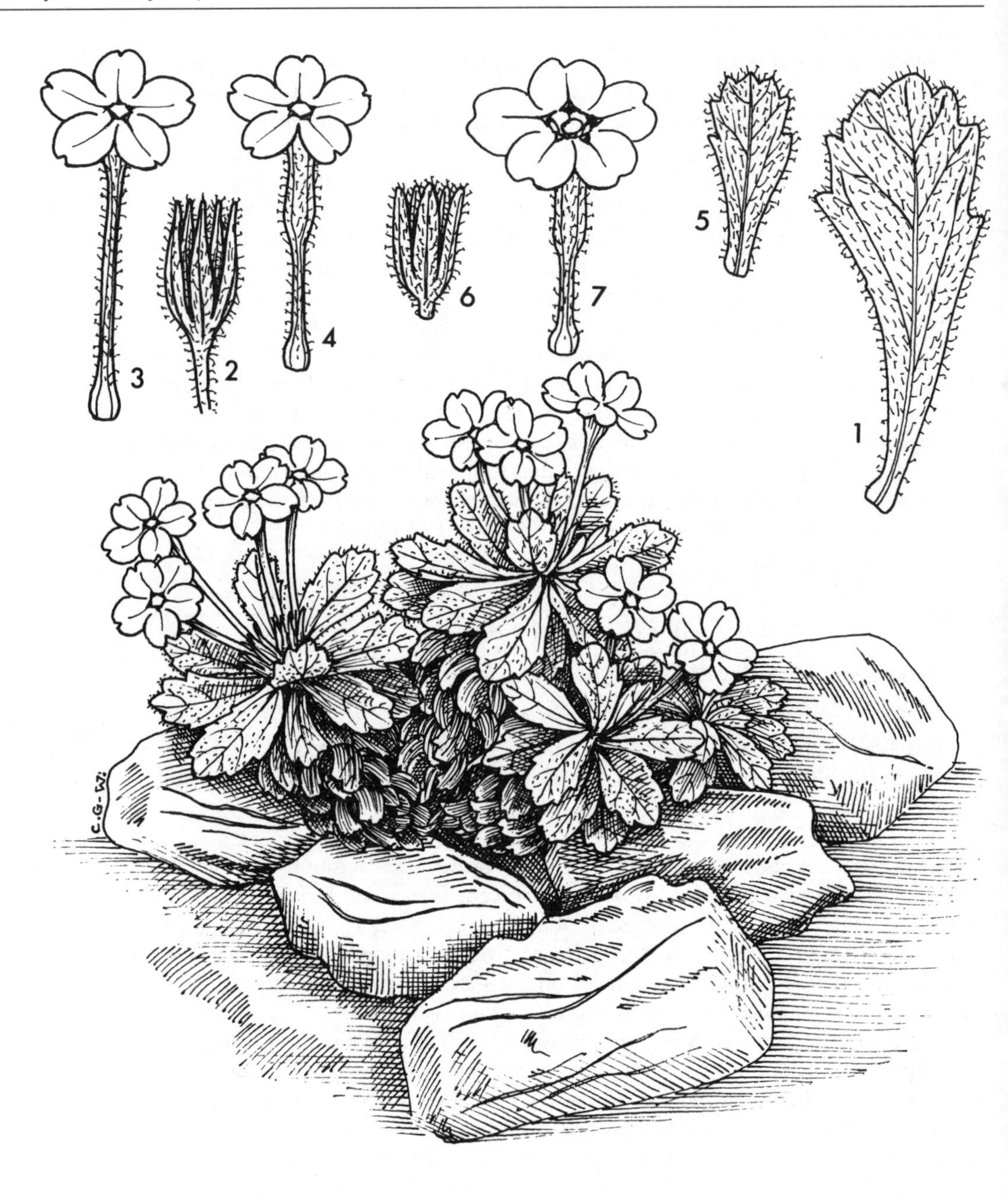

Dionysia lacei, × 1. 1 leaf, × 3; 2 calyx, × 3; 3 thrum-eyed corolla, × 2; 4 pin-eyed corolla, × 2. *D. saponacea*. 5 leaf, × 3; 6 calyx, × 3; 7 pin-eyed corolla, × 2.

is only locally abundant in the clefts of limestone rocks in shady situations at about 4500 ft.'

Clay (1937) transferred the species to *Dionysia* realising that the plant in question more properly belongs in that genus.

Dionysia lacei is the only species of *Dionysia* to be found in Pakistan. It is found in the mountains of Baluchistan close to the border of Afghanistan. The species probably finds its closest ally in *D. paradoxa* which is endemic to E Afghanistan. *D. paradoxa* has well developed scapes with the superposed whorls of flowers well spaced. In *D. lacei*, on the other hand, the scape is poorly developed and the whorls of flowers (often only a simple umbel in fact!) are crowded close to the leaf-rosettes. Although *D. lacei* is a smaller plant than *D. paradoxa*, with smaller leaves, the corollas are larger.

Dionysia lacei was discovered in 1888 by J. H. Lace. However, it was not collected again until 1965 when Jennifer Lammond (Edinburgh Botanic Garden) was on a field expedition to Baluchistan. It was subsquently also collected by Professor Rechinger from Vienna.

Seed was collected and introduced by Jennifer Lammond. However, the results were not particularly encouraging. Shortly after its introduction she wrote ... 'I had a certain amount of seed which was sent to Mr Esselmont in Aberdeen and Jim Archibald. There were a few plants with Mr Esselmont for a year or two, but I think they have now died'. Unfortunately we must wait for further introductions before we can try again, though there seems no reason why *D. lacei* should be any more tricky in cultivation than *D. balsamea* or *D. bornmuelleri.*

7 *Dionysia saponacea*

Dionysia saponacea Wendelbo & Rech. fil. in Årbok Univ. I Berg. Mat.-Nat. Ser. 19, IV: 10, fig. 3a-b (1963). Type: Afghanistan, Ghorat, Kuch-Tscheling-Safed-Daraq (Pirestan), July–August 1962, *Rechinger* 19092 (holotype W; isotype BG).

Description. *A tufted perennial* forming rather lax cushions, efarinose; branches becoming woody and bare below eventually, above with marcescent leaves and leaf-bases. *Leaves* forming lax rosettes, oblong to oblanceolate with a ± cuneate base, 15–30 mm long, 6–20 mm wide, including the petiole, the margin with 3–4 dentate or serrate-dentate teeth on each side, the lateral veins rather obscure, covered on both surface with articulated hairs and minute glands. *Inflorescence* reduced to a single flower; flowers heterostylous. *Peduncle* 1–1.5 mm long. *Bract* solitary, linear, c. 3 mm long, acute, glandular-pubescent. *Calyx* tubular-companulate, c. 6 mm long, divided for three-quarters of its length into ovate-lanceolate lobes. *Corolla* yellow, glandular outside (only the short-styled form is known); tube 18–20 mm long; limb 16–18 mm diameter, divided into obovate to suborbicular, slightly emarginate lobes. *Anthers* 2.5 mm long, inserted in the throat and ? in the middle of the corolla-tube. *Style* reaching about half-way along the corolla-tube or ? reaching the throat. *Capsule* with 7–10 seeds.

Distribution. WC Afghanistan, Ghorat Province, Kuh-Tscheling-Safed-

Daraq near Parjuman – not known elsewhere; altitude 2600–2800 m.

HABITAT. Limestone rock crevices, north-facing. Flowering probably during February and March.

Dionysia saponacea is closely related to *D. paradoxa*, indeed it can be likened to a reduced form of that species. The leaves are smaller and thicker, with fewer simple dentations and, most significantly, the inflorescence is reduced to a single practically scapeless flower. Because of the thickness of the leaves the lateral veins are rather obscure, whereas in *D. paradoxa* and its cousin, *D. balsamea*, they are far more prominent.

Dionysia saponacea is said to have a rather unpleasant 'antiseptic' smell. This is in contrast to the pleasant sweet spicy smell of *D. paradoxa* and *D. balsamea* and indeed most other species of *Dionysia*.

Dionysia saponacea has only been collected once. It was discovered on the Kuh-Tscheling-Safed-Daraq by Professor K. H. Rechinger of Vienna, which incidentally is also the type locality of *D. balsamea.* Despite the fact that these two species grow in close proximity to one another they are quite distinct.

Dionysia lacei, which belongs to the same association, like *D. saponacea* has a reduced inflorescence but the reduction is primarily in the scape, a number of flowers being bunched together amongst the leaf-rosettes. The leaves of *D. lacei* are about the same size as *D. saponacea*, but are thinner, more prominently veined, and are covered in short, curled, rather than straight, articulated hairs.

Dionysia saponacea is not in cultivation.

8 *Dionysia hissarica*

Dionysia hissarica Lipsky in Trudy Imp. St.-Petersb. Bot. Sada 18: 83 (1900) & in *loc. cit.* 23: 175, t.10 (1904). Type: USSR, Pamir Alai, above Den-Surkh, near river Surkhan, *Lipsky* s.n. (holotype LE; isotypes B, E, G).

Primula hissarica (Lipsky) Bornm. in Bull. Herb. Boiss. 2, 3: 592 (1903).

DESCRIPTION. *Plants* forming large loose tufts; branches becoming woody below, above with marcescent leaves and leaf-remains. *Leaves* borne in dense whorls on flowering shoots, but alternate on vegetative shoots, oblong to obovate or spathulate, 5–10 mm long, 2.5–5 wide, the margin with a few (3–5 on each side) coarse obtuse teeth, covered on both surfaces with a mixture of articulated hairs and minute glands, with rather dense yellowish woolly farina beneath; leaves of vegetative shoots generally larger, to 20 mm long. *Inflorescence* a simple 2–3-flowered umbel; flowers heterostylous. *Peduncle* 1–2.3 cm long, glandular-pubescent. *Bracts* foliaceous, oval to obovate, generally larger than the subtending leaves, pubescent like the leaves. *Calyx* campanulate, 6–8.5 mm long, divided practically to the base into lanceolate, generally somewhat toothed lobes, glandular-pubescent. *Corolla* yellow, glandular-pubescent outside; tube 13–15 mm long; limb 7.5–9 mm diameter, divided into ovate, entire lobes. *Anthers* c. 2–2.5 mm long, inserted in the throat or the middle of the corolla-tube. *Style* reaching the throat or the middle of the corolla-tube. *Capsule* with 4–10 seeds.

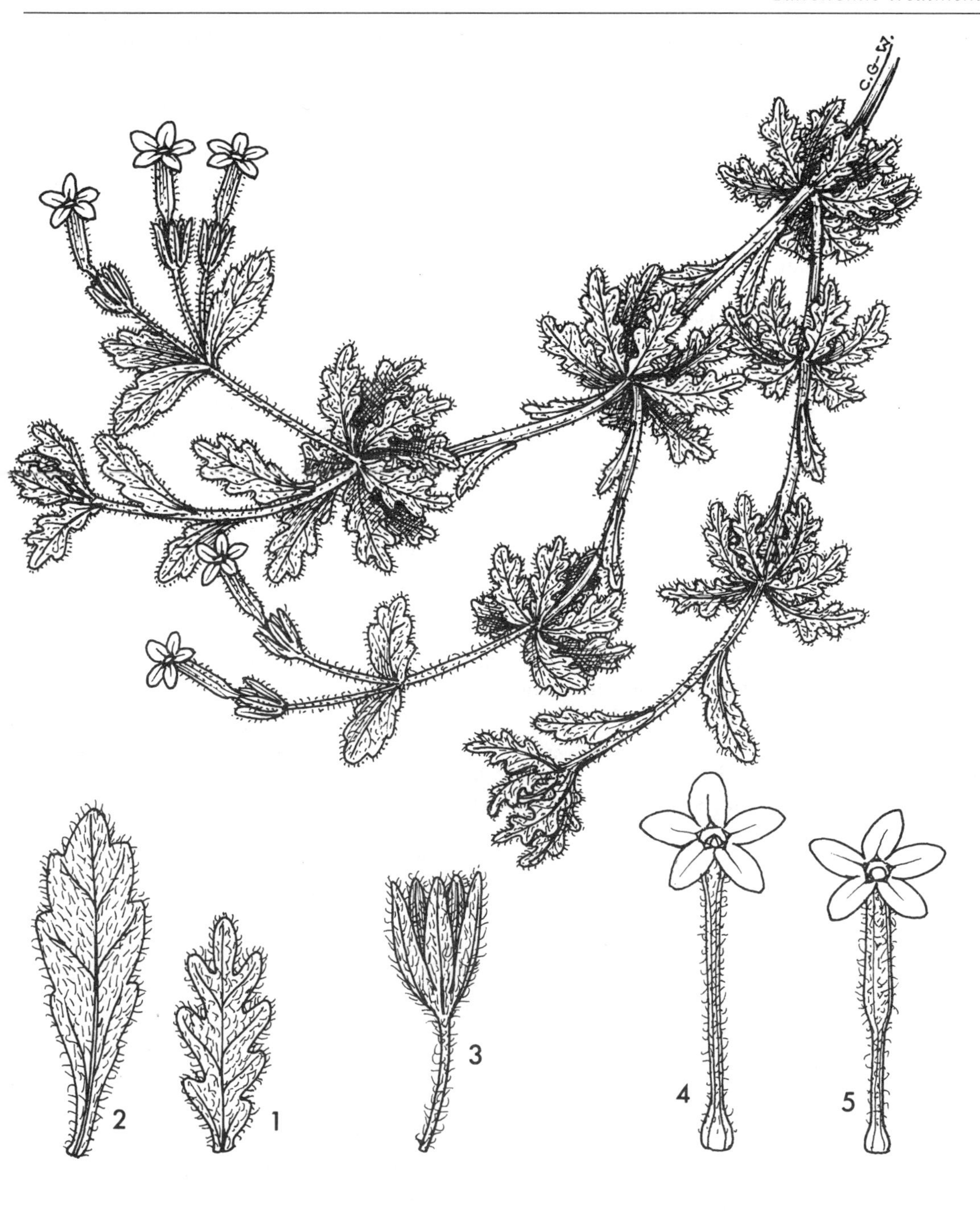

Dionysia hissarica, × 1. 1 leaf, × 2; 2 bract, × 2; 3 calyx, × 3; 4 corolla, × 3; 5 pin-eyed corolla, × 3.

DISTRIBUTION. S USSR; Pamir Alai, Hissar region, Den-Surkh and Khondiz R. – not known elsewhere; altitude c. 1600 m.

HABITAT. North-facing damp sandstone cliffs. Flowering during April and May.

Dionysia hissarica was first discovered and collected by Lipsky in 1896 in the Pamir Mountains of S USSR. For many years after little was known of this species, indeed its precise locality was unknown. However, in 1967 it was recollected by L. A. Smolianinova, on the left bank of the Khondiz river, a tributory of the Sangandak in the Pamir Alai, not far from Dibadam. Seedlings were raised at the Leningrad Botanic Garden and the chromsome count of 2n = 20 ascertained. It is not known whether it is still in cultivation in Leningrad.

Dionysia hissarica is an extraordinary and distinct species, unmistakable with the leaves of flowering shoots arranged in dense whorls and with few-flowered umbels of yellow flowers. In contrast, the long vegetative shoots bear alternate leaves, but eventually terminating in a whorl of leaves that will support the following year's inflorescence.

This interesting species holds a rather isolated position within subsection *Scaposae*, but presumably finds its closest allies in those species of the subsection native to Afghanistan – *D. paradoxa* and *D. balsamea* for instance. However, Wendelbo (1961) places it closest to *D. bornmuelleri*, although he also says that – 'In seed characters it comes closer to *D. paradoxa* which probably is its closest kin'.

Early drawings of *D. hissarica* were misleading. Wendelbo (1961) has amply clarified this – 'It has been considered as somewhat anomalous within the genus because of the pendant flowers shown in Lipsky's drawing (1904, t.10) and reproduced by Smolianinova (1952, t.XI). After having studied the sheet of the Berlin herbarium which is quite adequate material, I have come to the conclusion that Lipsky's drawing is very misleading. As *D. hissarica* most probably grows in crevices of steep rocks and the shoots hang down, the flowers will be erect . . . not pendant as Lipsky's drawing suggests. There is now nothing anomalous about it'.

Despite this, *D. hissarica* remains something of an enigma. Why it has not been collected more frequently it is difficult to say, for it grows in the W Pamir Mountains, and although the precise type locality is uncertain, it cannot be very far from the localities of that other Pamir endemic, *D. involucrata*. Perhaps further botanical exploration of this remarkable mountain range, which lies close to both the Hindu Kush and the Karakoram, will reveal further localities for this plant.

9 *Dionysia aretioides*

Dionysia aretioides (Lehm.) Boissier, Diagn. Plant. Orient. Nov. 1(7): 68(1846). Type: Iran, Ghilan, *Hablitzl* s.n. (holotype ?LE; isotypes B, BN, C, G, W).

Primula aretioides Lehmann, Monographia generis Primularum: 90, t.9 (1817).

Primula cespitosa Willd. ex Roemer et Schulter, Syst. Veg. 4: 785 (1819), *pro syn.*

Primula caespitosa Willd. ex Steudel, Nomen. Bot.: 655 (1821), *pro syn.*

33 *Dionysia bryoides* growing on limestone cliffs, Kuh-i-Dinar; being pollinated by a fritillary butterfly. S. Iran. Photo. T. F. Hewer.

34 *Dionysia bryoides*. Photo. E. G. Watson.

35 *Dionysia zagrica*, growing on limestone cliffs, Kuh-i-Sehquta, SW Iran, the type locality – just past flowering. Photo. T. F. Hewer.

36 *Dionysia denticulata* growing on limestone cliffs, NW of Ghazni, CE Afghanistan. Photo. C. Grey-Wilson.

37 *Dionysia denticulata* growing on limestone cliffs, NW of Ghazni, CE Afghanistan. Photo. C. Grey-Wilson.

38 Limestone cliffs of the Doab Gorge, C. Afghanistan, where *D. tapetodes* grows in abundance. Photo. C. Grey-Wilson.

39 Limestone gorge N. of the Unai Pass, C. Afghanistan, on whose cliffs *D. tapetodes* can be found. Photo. C. Grey-Wilson.

40 Cushions of *D. tapetodes* (yellowish green) and *D. viscidula* (dark green, on left) on limestone cliffs, Darrah Zang, Afghanistan. Photo. C. Grey-Wilson.

Gregoria aretioides (Lehm.) Duby in DC. Prodr. 8: 46 (1844).

Dionysia aretioides var. *typica* Knuth in Pax & Knuth, Pflanzenreich 4 (237): 165, fig. 141 (1905).

Dionysia demawendica Bornm., Plantae Brunsianae in Beih. Bot. Centralbl. 33 (2): 301, t.2 fig. 1 (1915). Type: N. Iran, Mt. Demawend, Abigerm, *Bruns* s.n. (holotype B).

DESCRIPTION. *Plants* forming loose to rather dense greyish cushions (up to 40 cm diameter in the wild), farinose or efarinose. *Stems* much-branched, becoming woody and bare below, covered in spreading dead leaves and leaf-remains and terminating in leaf-rosettes. *Leaves* linear-oblong to ± spathulate, 5–7 mm long, 1–1.5 mm wide, the apex obtuse, the margin strongly revolute, with 4–5 blunt teeth on each side in the upper half, covered in long articulate hairs and sometimes with minute capitate glands, often with white to yellowish woolly farina below; leaves of vegetative, non-flowering shoots larger, to 12 mm long and 7 mm wide, with spreading, scarcely revolute margins; veins on all leaves relatively inconspicuous. *Flowers* generally solitary, sometimes 2 borne on very short, 1–2 mm long, peduncles. *Bracts* 2, unequal, linear-lanceolate, 4–8 mm long, entire, pubescent like the leaves. *Calyx* campanulate, 5–7 mm long, divided for above three-quarters of its length into linear-lanceolate lobes, covered in long articulate hairs. *Corolla* yellow, pubescent outside, heterostylous; tube 13–18 mm long; limb 8–10 mm diameter, divided into broad- to narrow-obcordate lobes, shallowly notched. *Anthers* 2–2.5 mm long, inserted in the throat or in the middle of the corolla-tube. *Style* reaching just below the throat or the middle of the corolla-tube. *Capsule* with 20–30 seeds (Plates 10–13).

DISTRIBUTION. N Iran: C Elburz Mountains, particularly on the Caspian side, especially the Chalus and Haraz Valleys, the Kandevan Pass and the vicinity of the mountains Demawend and Takht-e-Suleiman; altitude 300–3200 m.

HABITAT. Growing primarily on shaded (often moist in spring) limestone cliffs, north- west- or east-facing, on rock ledges and in deep crevices. Flowering in March and April.

Dionysia aretioides was the first species in the genus to be discovered. It was first collected by Hablitzl in 1770 on Samamys Kuh in the Elburz Mountains of northern Iran. The species was first described some years later by Lehmann in 1817 who placed it in the genus *Primula* – *Primula aretioides* Lehmann, but it was not transferred to *Dionysia* until 1846 when Boissier published his *Diagnoses Plantarum Orientalium Novarum*.

Dionysia aretioides is endemic to the Elburz mountains of northern Iran, which flank the southern shore of the Caspian Sea. This species is mainly found along the limestone cliffs of various valleys leading over the mountains from the south, but it is found quite frequently along the drier parts of the gorges of the northern flanks; the northern flanks are generally moister because of the proximity of the Caspian Sea. No other species of *Dionysia* is known from the Elburz Mountains.

In many ways it is curious that little further information on this species was

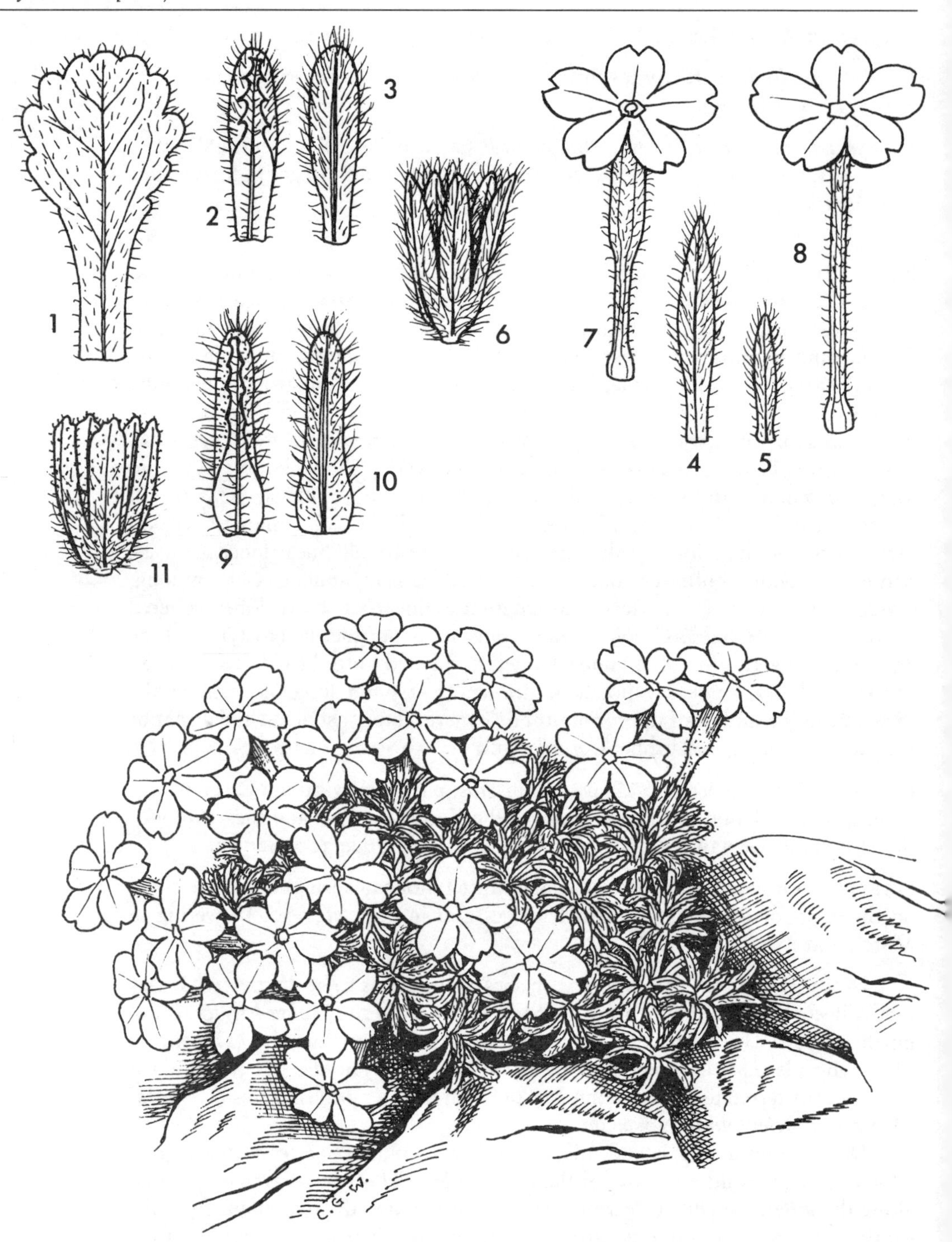

Dionysia aretioides, × 2. 1 leaf of vegetative shoot, × 4½; 2 leaf below, × 4½; 3 leaf above, × 4½; 4–5 bracts, × 6; 6 calyx, × 4½; 7 pin-eyed corolla, × 3; 8 thrum-eyed corolla, × 3. *D. leucotricha*. 9 leaf below, × 4½; 10 leaf above, × 4½; 11 calyx, × 4½.

known until 1942 when Bornmüller published a full diagnosis based on material collected by Gauba. One would have expected more information before then as the Elburz was for many years the most thoroughly explored area botanically in the whole of Iran.

It is therefore even more surprising that this species did not come into cultivation until 1959 from seed collected by Professor Per Wendelbo – the first plants flowered in 1961. The first plants to flower in Britain were, however, reared from seed collected by Paul Furse in 1961 and since then there have been several subsequent collections.

In the wild two forms of *D. aretioides* can be recognised, a farinose one and an efarinose one. Of the two, the former is by far the commonest and all plants in cultivation have farina to some extent.

Dionysia aretioides finds its closest ally in *D. leucotricha*, indeed Bornmüller considered them to be one and the same species for a long time. The main differences are to be found in the calyx; in *D. leucotricha* the upper part of the calyx lobes are hyaline, almost invariable toothed and scarcely hairy – the hairs being few and short. The leaves of *D. leucotricha* are aggregated into more distant glomerules (each representing one years growth) whilst they are crowded together in *D. aretioides*. In addition the leaves and calyces of *D. leucotricha* have numerous small capitate glands mixed with the longer hairs, whereas such glands are few in *D. aretioides*. The seed capsules of *D. leucotricha* also contain fewer seeds. *D. leucotricha* occurs in a number of localities in the northern Zagros Mountains of Iran, well isolated from *D. aretioides*. These two species are undoubtedly very closely allied but differ in a sufficient number of details to warrant inclusion here as distinct species. Unfortunately *D. leucotricha* is not in cultivation, so that a direct comparison of living material is not yet possible.

Dionysia aretioides is one of the easiest and most accommodating species in cultivation, fast growing and often very floriferous. Plants can make large cushions and in five to seven years they may measure 25–32 cm in diameter. Plants eventually get too large and have to be replaced. This is not a major problem as new young plants can be easily propagated and young floriferous plants are very attractive. Jack Elliott wrote in 1969 'I have a dozen large plants grown from original PF seed, and they present no difficulty but vary considerably in floriferousness. It seems to me that they like a fairly rich soil and plenty of moisture. They are easily propagated from cuttings'.

That this species likes plenty of moisture was emphasized at the same time by Peter Edwards – 'This plant has now been in successful cultivation for a number of years and the form which I grow seems very free-flowering ... water in plenty during the growing period and also in the winter from time to time, as a number of cultivators, I am sure, have lost plants through winter drought'.

This fine species is quite variable in cultivation and a number of clones, several named, are available. Unfortunately some of these have been muddled and their origins obscured and one should seek named clones from a reliable source. Eric Watson reports that – 'I think the various clones are now mixed up. I've seen the clone 'Paul Furse' on show benches with both pin and thrum and the same with 'Gravetye'.' These are the best-known clones; both fast growing, very floriferous and forming rather large loose cushions; the flowers are relatively large with broad petal lobes compared to some of the unnamed clones. A clone referred to as 'BBF' has according to Chris Norton – 'Bloody big flowers! We'll have to find a better name for this form'. 'Phyllis Carter' is perhaps the most charming cultivar of all with its tighter, rather smaller cushions and neat flowers. All the named clones must be propagated from cuttings otherwise their authenticity cannot be guaranteed.

In the *Alpine Garden Society Bulletin* (37: 349 1969) Doris Saunders comments that ''Paul Furse' is the second plant of *D. aretioides* to receive an Award of Merit under a clonal name, the first being 'Gravetye' in 1966. Admittedly, both these plants are outstandingly good and well worthy of the honour they received. But *D. aretioides* is such a variable species and exhibits so many fine forms, each one differing from the next both in shape, size and colouring of its flowers and in the appearance of its foliage, that it seems invidious to confine any award to those few specimens which are able to be brought before a committee, instead of honouring the species as a whole for being a quite outstanding one'

Most collections generally contain both pin- and thrum-eyed plants and seed is generally produced naturally without having to resort to hand-pollination. However, if the weather is overcast and damp when the plants are in full bloom then some hand-pollination is probably desirable.

In most instances, though, plants are best raised from cuttings. These are readily rooted and under normal conditions one can expect a success rate of 70–90%. Seeds, on the other hand, generally show a low percentage germination, usually germinating after the first winter.

Dionysia aretioides is perhaps the most accommodating species of *Dionysia* in cultivation. It will succeed in a variety of well-drained composts and will thrive on tufa. Several growers have grown it very successfully out-of-doors in a trough or on a raised scree where the plants are given the protection of a sheet of glass or cloche during the winter months. Roy Elliott had an established plant on his tufa cliff for a number of years and the author kept one going in a trough outdoors for five years where it flowered annually and was protected solely by the overhanging eaves of the house – the plant eventually succumbed to the very low temperatures and prolonged freeze of the winter of 1985–86.

Diseases and pests do not appear to be a problem, although plants are occasionally attacked by aphids, particularly in the early spring.

There is no doubt that *D. aretioides* is a very fine species but its very success and ease of cultivation can pose a problem, as Eric Watson points out – 'I don't grow this species in a pot any more – can't keep up with it as it wants repotting about twice a year. Very greedy.'

Nevertheless *D. aretioides* is the ideal species for the novice grower to start with – plants are relatively readily available and not too expensive! Some forms have sweetly scented flowers.

Artificial hybrids have been produced between *D. aretioides* and *D. bornmuelleri* (p. 167), *D. mira* (p. 167) and *D. teucrioides* (p. 167).

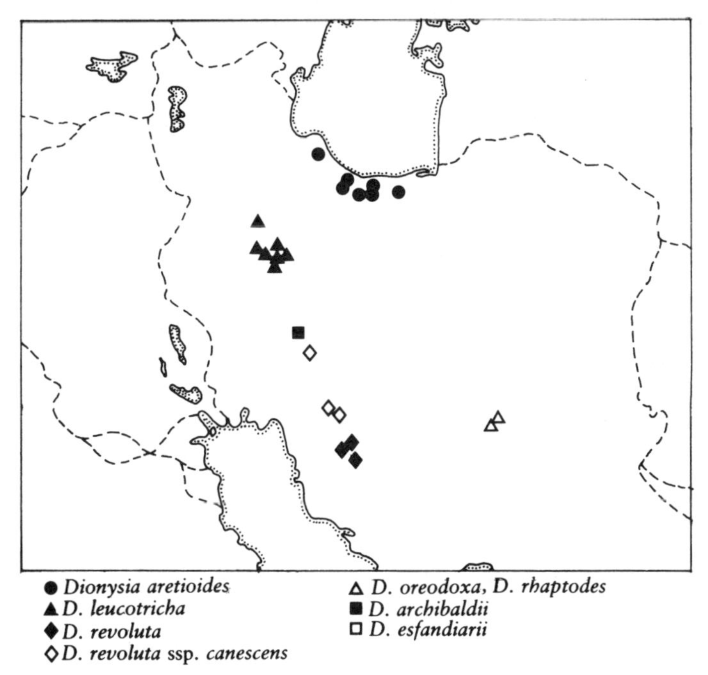

Map 3 Distribution of section *Anacamptophyllum* subsect. *Revolutae.*

10 *Dionysia leucotricha*

Dionysia leucotricha Bornm., Coll. Straussianae nov., Beih. Bot. Centralbl.28(2): 460 (1911). Type: Iran, *Strauss* (syntypes B,JE).

Dionysia aretioides (Lehm.) Boiss. var. *adenophora* Bornm. in Bull. Herb. Boiss. ser. 2, 3: 593, t.6 fig. 5 (1903). Type: as above.

DESCRIPTION. *Plants* forming large to fairly dense greyish cushions (to 30 cm diameter in the wild). *Stems* bare and woody below, above covered with spreading leaves and leaf-remains generally in discrete whorls, terminating in leaf-rosettes. *Leaves* oblong to narrow-obovate, 5–7 mm long, 1–1.5 mm wide, the apex obtuse, the margin revolute and with about 4 blunt teeth on each side in the upper two thirds, covered on both sides by long articulate hairs and minute capitate glands, often with white woolly farina beneath; veins rather obscure, but midrib distinctly raised beneath. *Flowers* solitary, rarely 2, borne on a very short 1–3.5 mm long pedicel. *Bracts* 2, unequal, linear-lanceolate, pubescent like the leaves. *Calyx* campanulate, 4.5–5.5 cm long, divided for two-thirds of its length into oblong-lanceolate lobes, often slightly toothed near the top, covered at the base in articulate hairs and minute capitate glands, but subglabrous in the upper half. *Corolla* yellow, glandular-pubescent outside, heterostylous; tube 14 mm long; limb 9–12 mm diameter, divided into obcordate, shallowly emarginate lobes. *Anthers* 1.5 mm long, inserted in the throat or just above the middle of the corolla-tube. *Style* reaching the throat or one third the way up the corolla-tube. *Capsule* with 15–20 seeds.

DISTRIBUTION. W. Iran; mountains in the vicinity of Hamadan, Nehavend, Burujird, Sultanabad and Khorramabad; altitude ? m.

HABITAT. Shaded and semi-shaded limestone rock crevices. Flowering March–April in the wild.

Dionysia leucotricha was described in 1903 as var. *adenophora* of *D. aretioides* by Bornmüller. However, in 1911 he raised the variety to specific status giving it the name *D. leucotricha*. As has already been discussed, *D. leucotricha* (p 79) bears a close resemblance to *D. aretioides*, but differs in a number of characters.

Nearly all the known collections of *D. leucotricha* were made by Th. Strauss between 1902 and 1904. Strauss made numerous collections of Dionysias in Iran during this period. What is remarkable is that little has been seen of the species since then. Judging by Strauss's collections the species occurs in a number of localities at the northern end of the Zagros Mountains close to Hamadan and Sultanabad, an area apparently little explored by modern plant-collectors. As a result *D. leucotricha* has never been in cultivation, but it would almost certainly be a fine addition and we must await more enlightened times before we can see living material.

In the wild it almost certainly inhabits drier regions that *D. aretioides* and it is likely to prove rather more exacting in cultivation.

For a discussion of the differences with *D. aretioides* see p. 79.

11 *Dionysia revoluta*

Dionysia revoluta Boiss. Diagn. Plant. Orient. Nov. 1(7): 65 (1846).
Type: Iran, Fars, Kuh-i-Sabzpuchon (Sabz Puchon), *Kotschy* 426 (holotype G; isotypes B, BM, K, W.).

Dionysia revoluta var. *typica* Knuth in Pax & Knuth, Pflanzenreich 4(237): 161, fig. 40a–f (1905).

Primula revoluta (Boiss.) Bornm. in Bull. Herb. Boiss. 7: 73 (1899).

DESCRIPTION. *A laxly branched subshrub* to 70 cm across, though generally less; branches becoming woody and bare below, above with marcescent spreading or reflexed leaves. *Leaves* forming small lax rosettes, those of flowering shoots narrow-oblong or elliptic or linear, the margin revolute with 7–8 small obtuse teeth on each side, the midrib raised beneath, covered in minute capitate glands and sparse or dense articulated hairs, often with whitish or yellowish woolly farina beneath; leaves of vegetative shoots larger, to 18 mm long and 6 mm wide, flat-margined. *Flowers* 2–4 together, or solitary, subsessile, heterostylous. *Bracts* solitary linear, c. 5 mm long, glandular. *Calyx* tubular, 5–6 mm long, divided halfway into narrow-triangular, acute lobes, glandular-pubescent. *Corolla* yellow, glandular-pubescent outside; tube 15–18 mm long; limb 10–12 mm diameter, divided into broadly obcordate lobes, deeply notched.

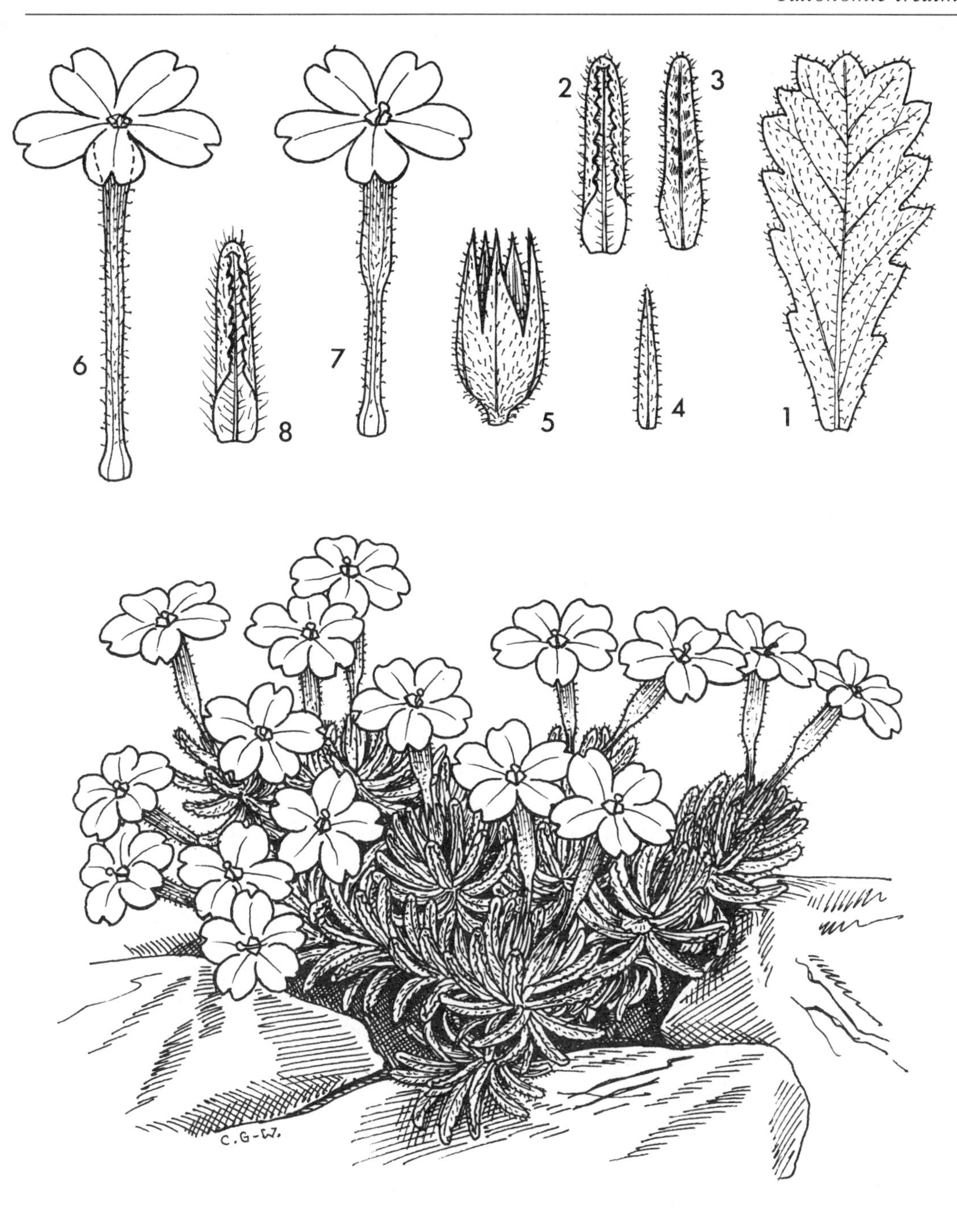

Dionysia revoluta subsp. *revoluta*, × 2. 1 vegetative leaf, × 4½; 2 leaf below, × 4½; 5 calyx, × 4½; 6 thrum-eyed corolla, × 3; 7 pin-eyed corolla, × 3. Subsp. *canescens*; leaf below, × 4½.

Anthers c. 2 mm long, inserted in the middle of the corolla-tube or in the throat. *Style* reaching the middle of the corolla-tube or slightly exserted. *Capsule* containing 20–30 seeds.

subsp. **revoluta**

Plants deep green. *Leaves* with a rather sparse covering of short articulate hairs, not more than 0.3 mm long, occasionally subglabrous, except for the leaf-base and mid-rib beneath (Plate 14–15).

DISTRIBUTION. SW Iran: Fars Province – Kuh-i-Sabzpuchon, Kuh Akhbar Ali, Kotel Doun and various mountains close to Shiraz; altitude 1800–3000 m.

HABITAT. Shady limestone cliffs, particularly below overhangs. Flowering from late March to early May.

subsp. **canescens** (Boiss.) Wendelbo in Årbok Univ. I Bergen. Mat.-Nat. Ser. 3,1: 48 (1961).

var. *canescens* Boiss., Fl. Orient. 4: 18 (1879). Type: Iran, Luristan, Kuh Eschker, *Haussknecht* s.n. (lectotype G).

Plants usually greyish or grey-green. *Leaves* densely covered in long articulated hairs, mostly at least 0.4 mm long (Plate 16).

DISTRIBUTION. CW & SW Iran: Fars Province – Kuh-i-Daena (= Dinar), vicinity of Yasuj and Sisakht, Kuh-i-Sehquta; Luristan Province – Kuh Eschker, Tang Serastane, Kuh Maregun, Mt Kellar by Sebze river; altitude 2000–2900 m.

HABITAT. Limestone cliffs in shaded or semi-shaded places, rarely in full sun. Flowering April–May.

Dionysia revoluta was first collected in 1842 by Kotschy, just south of Shiraz in SW Iran. Since then it has been re-collected on a number of occasions on the Kuh-i-Sabzpuchon (= Imam Zade Sabzpuchon) and on various other mountains in the Shiraz area, but also in Luristan Province in the Bakhtiari Mountains considerably further north. More recently, however, a number of collections by Tom Hewer (in 1973) at intermediate stations in the S Zagros Mountains around Kuh-i-Dinar, have partly bridged the gap between the northern and southern localities. It must certainly grow at other localities in these mountains, where suitable cliffs can be found.

On the massive cliffs of the Kuh-i-Sabzpuchon *D. revoluta* often forms substantial bushes, sometimes inhabiting overhanging limestone rocks where water drips down. In other, drier places, on these same cliffs it can be found in association with *D. bryoides* and *D. diapensiifolia*.

Dionysia revoluta is perhaps more bushy than any other species of *Dionysia*. Older plants have a twiggy and much-branched appearance, the stems woody and bare below. The species clearly finds its closest allies with the other members of subsection *Revolutae*. With its neatly toothed revolute leaf-margins, yellow flowers and notched petals it clearly has its closest affinities with the northern Iranian *D. aretioides* and *D. leucotricha*. However, it can be quickly distinguished by its looser, twiggy rather than compact growth, the leaf-margin with 7–8 rather

41 *Dionysia tapetodes* growing on a limestone cliff. Andaräb Valley, E. of Banu, E. Afghanistan. Photo. C. Grey-Wilson.

42 *Dionysia tapetodes*. Photo. S. Taylor.

43 *Dionysia lindbergii*. Photo. S. Taylor.

44 Limestone cliffs of the Varzob Gorge, Pamir Alai, S. USSR, the home of *D. involucrata*. Photo. R. B. Burbidge.

45 *Dionysia involucrata*. Photo. S. Taylor.

than 4 pairs of teeth and the less deeply divided calyx.

Subsp. *canescens* was described from Kuh Eschker in Luristan Province. Plants are distinguished solely by the denser covering of longer and more prominent articulated hairs on the leaves, bracts and calyces, which give the whole plant a much greyer appearance. These hairs have a far denser underlayer of capitate glands than in the typical subspecies (subsp. *revoluta*). Grey-Wilson (1974) has pointed out that specimens collected at intermediate localities by Tom Hewer, although placed in subsp. *canescens*, in some measure bridge the gap between subsp. *revoluta* and subsp. *canescens* and that other localities in the region may well reveal further transitions. In that eventuality it may not be practical to uphold the two subspecies. In the meantime, however, there is just enough evidence to separate plants from the environs of Shiraz with all the material recorded to date from further north.

Dionysia revoluta was introduced into cultivation in 1966 by Jim Archibald as seed, and flowered for the first time 3 years later.

Peter Edwards wrote in 1969 – 'A shrub-like species, eventually forming a large loose cushion similar to *D. aretioides* and flowering at the same time, with the flowers somewhat similar. As with *D. aretioides* the plant is a strong grower and needs regular repotting – it can be comfortably knocked from its pot for this purpose. Again ample water is required during the growing season and just moisten round the rim of the pot in winter. Propagation by green cuttings in early spring, or half-ripe cuttings in late autumn ... The species seems to be host to greenfly to an even greater extent than *D. caespitosa* ... therefore regular spraying with an insecticide is essential'.

Cuttings give a success rate of between 40% and 90%, the cuttings rooting generally in 6–8 weeks, sometimes longer. Stem-cuttings give the best results, but leaf-cuttings have given some success, although they are more prone to botrytis.

Mature plants are reasonably hardy but winter temperatures anywhere below – 8°C may damage or kill plants, even those kept reasonably dry. Some protection in the alpine house during periods of severe weather is advisable.

Brian Burrow reports growing a plant outside on a raised scree bed, but giving winter protection from November until February. However, it succumbed to severe winter weather after two seasons success.

Several growers have had great success growing plants under cover in tufa, plants retaining a more compact habit; plants tend to become loose and rather leggy with age. They may reach 15 cm across in 4–5 years after which they generally require replacing by propagation. Stan Taylor, however, has a 12 year old plant that has reached 33 cm across, certainly the largest in cultivation.

Most of the plants (but not all) are referable to *D. revoluta* subsp. *canescens* from seed collected by Professor Hewer in Iran in 1973 – numbers H1949 and H1980,

Plants flower from February to April in cultivation.

12 *Dionysia archibaldii*

Dionysia archibaldii Wendelbo in Bot. Not. 120: 144 (1967), figs. 2–3.
Type: Iran, Bakhtiari, Tang-i-Sirdan, Aug. 1966, *J. Archibald* 3053 (holotype GB).

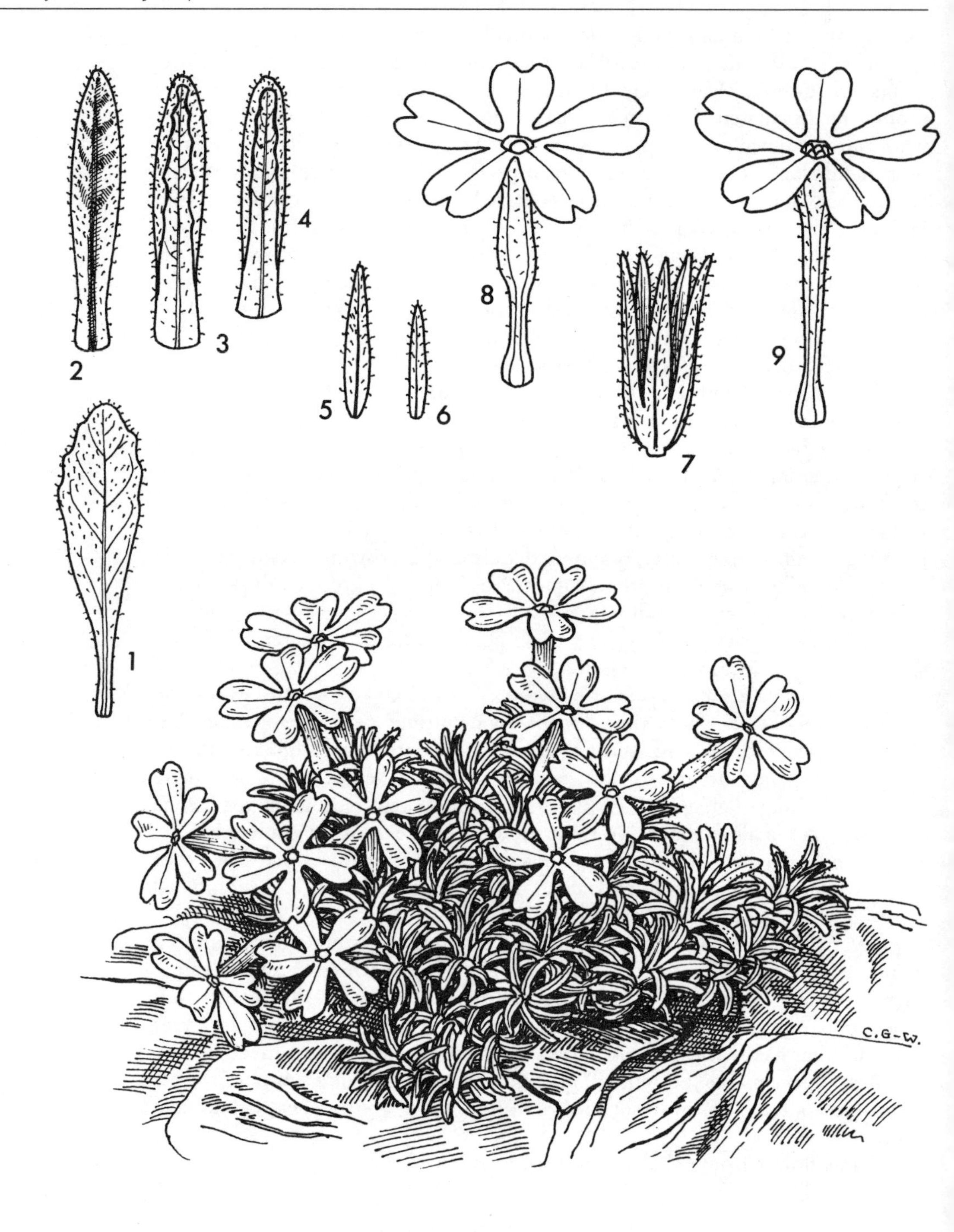

Dionysia archibaldii, × 1½. 1 vegetative leaf, × 6; 2 leaf above, × 6; 3–4 leaves below, × 6; 5–6 bracts, × 6; 7 calyx, × 6; 8 pin-eyed corolla, × 3; 9 thrum-eyed corolla, × 3.

DESCRIPTION. *Plants* forming a rather loose greyish cushion (to 20 cm diameter in the wild), farinose. *Stems* thin, becoming woody below eventually, with the dead leaves and leaf-remains forming distinct whorls, closely overlapping within each whorl. *Leaves* oblong-elliptic to oblong, 4–7.5 mm long, 1–1.5 mm wide, the apex subobtuse, the margin distinctly revolute, the unrolled margin ± entire or crenate with 3–4 shallow teeth on each side, covered in short glandular hairs; leaves of young plants often larger, to 10 mm long; veins on all leaves rather inconspicuous. *Flowers* solitary, sessile. *Bracts* 2, linear-lanceolate, subacute, 3–4 mm long, glandular like the leaves. *Calyx* tubular-campanulate, 4–5 mm long, divided for three-quarters of its length into linear-lanceolate, entire lobes, glandular-pubescent. *Corolla* violet, glabrous to slightly glandular-pubescent; tube 12–14 mm long; limb 12–16 mm diameter, the lobes narrow-obcordate, deeply emarginate, almost bilobed. *Anthers* c. 1.2 mm long, inserted in the throat or just below the middle of the corolla-tube. *Style* reaching one-third or two-thirds of the way along the corolla-tube. *Capsule* containing 5–10 seeds (Plate 17).

DISTRIBUTION. W Iran: Bakhtiari; Zagros Mountains, between Kuhrang and Bazuft Valleys and Zardeh Kuh.

HABITAT. Shaded and semi-shaded limestone cliffs, in narrow crevices; altitude 2600–4300 m. Flowering in the wild in May and June.

Dionysia archibaldii is one of the most beautiful members of the *Revolutae*,

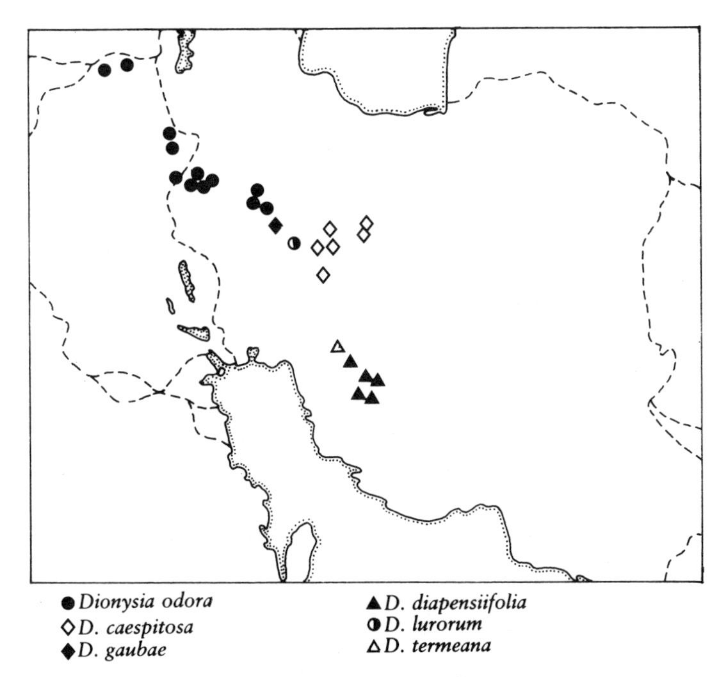

Map 4 Distribution of section *Dionysia* subsect. *Caespitosae*.

distinct on account of its violet-coloured flowers. The only other member of the *Revolutae* with violet flowers is the recently described and little-known *D. esfandiarii* Wendelbo.

Dionysia archibaldii was first discovered in 1966 by Jim Archibald in the Zagros Mountains, Bakhtiari Province. To date it has only been found in several close localities centred upon Zardeh Kuh and Tang-i-Sirdan, both well to the west of the city of Esfahan. Archibald found plants growing on limestone cliffs of north and north-west exposure at a high altitude, 4100–4300 m. However, in late May and early June 1976 Professor Hewer recollected this species at a considerably lower altitude, 2600–2800 m, in the Zardeh Kuh Gorge, where it inhabited 'shaded limestone or conglomerate cliffs'.

As in several other species, *D. archibaldii* appears to exist in the wild in both a farinose and an efarinose form. The early collections showed the leaves to have entire or slightly crenulate revolute margins, but the later collections of Hewer often have rather larger leaves, the margins clearly toothed, with 3–4 small teeth on either side. This character shows *D. archibaldii* to have very close affinities with *D. revoluta*, though of course the latter has bright yellow flowers. In the original description, Wendelbo states that the corolla-limb is about 10 mm in diameter, however, the later collections have clearly larger flowers with a corolla diameter of 15–16 mm.

From *D. esfandiarii* it differs in being generally larger in all its parts, less densely tufted and with deeply bifid corolla-lobes. In *D. esfandiarii* only the distal part of the leaf-margin is revolute and then always entire. The pubescence of these two species is also markedly different; the foliage, bracts and calyces of *D. archibaldii* are covered in shortly stipitate capitate glands, whereas those of *D. esfandiarii* have short ascending articulated hairs in the lower half, but with long spreading hairs above.

The first plants of *D. archibaldii* to flower in cultivation were raised from the 1966 Archibald collections. The first flowered in 1969 and at that time the late Peter Edwards wrote – 'I have two forms: a non-farinose and a farinose form ... the first makes a caespitose cushion, whilst the latter is not unlike *D. revoluta*, being more shrub like.

Propagation is by green cuttings in late April, May or early June, in sand in a north-facing cool position. I find this species one of the more difficult to strike...

Very little watering is required in the winter if plunged into not-too-moist material, but water in plenty during the growing season...'

Dionysia archibaldii is a rather pale-flowered species, but pretty enough. Plants form lax cushions which soon become rather straggly. Unfortunately it is not particularly easy to keep going for long and rarely exceeds four years old – plants are at their best in the second and third years. Four year old cushions may reach 6–8 cm diameter; the largest recorded plant being 10 cm.

All the original plants in cultivation came from Archibald collections of 1966, but some seed has been produced in cultivation and a number of clones are in existence. It has to be said that none are particularly floriferous. No grower recommends growing *D. archibaldii* in tufa. However, Brian Burrow managed to keep a single plant growing outdoors on a raised scree bed covered between November and February.

Fortunately this species is quite readily propagated from cuttings, often giving one hundred per cent success. Jack Wilkinson reports that – 'cuttings strike very well in 5–6 weeks in sand, under a hood, in shade – taken in May'. Cuttings can be rooted with varying degrees of success any time between May and August.

Both pin- and thrum-eyed plants are in cultivation so that it should be possible to produce a regular, if small, amount of seed.

The chief danger to older plants is botrytis which may infect the dead (marcescent) foliage and signs of such an attack should be carefully monitored.

Perhaps suprisingly, plants may produce a second crop of flowers in the autumn, but as Stan Taylor remarks – 'In my experience if this species flowers in the autumn then it always dies'.

Dionysia archibaldii exists in the wild in both a farinose and an efarinose form. Eric Watson points out – 'When Archibald first discovered this species he introduced two forms, farinose and efarinose. The efarinose was quickly lost and we were left with one clone which was kept going by cuttings.

Several years ago one of my plants set seed and I raised a batch of plants. All but one were efarinose. This leads me to think that the efarinose form is probably plentiful in the wild.

I now have several clones – pins and thrums, and I now get seed regularly'.

13 *Dionysia rhaptodes*

Dionysia rhaptodes Bunge in Bull. Ac. Imp. Sci. St.-Petersb. 16: 562 (1871). Type: Iran, Kerman, between Chabbis and Kerman, *Bunge* s.n. (holotype LE; isotypes G, K, P).

D. heterochroa Bornm. in Bull. Herb. Boiss. 7: 72, t.2 fig. 3 (1899). Type: Kuh-i-Jupar, *Bornmüller* 3872 (holotype B; isotypes G, JE).

Primula heterochroa (Bornm.) Bornm., *loc. cit.*

Description. *Plants* forming dense greyish tufts, often farinose. *Stems* short-branched, densely covered for most of their length by dead, partly overlapping leaves and leaf-remains, terminating in a dense leaf-rosette. *Leaves* ovate-oblong to lanceolate-oblong, 2.5–3.5 mm long, 0.8–1.5 mm wide, the apex obtuse, the margin entire, revolute in the upper half, sparsely to densely covered in short articulate hairs, rarely ± glabrous, often with white woolly farina below and/or in the leaf-axils; midrib raised beneath. *Flowers* solitary, very occasionally 2, sessile. *Bracts* 2, equal, linear-oblong, 3.5 mm long, pubescent like the leaves. *Calyx* tubular-campanulate, 3.5–4 mm long, divided almost to the base into linear oblanceolate, subobtuse lobes, covered in short articulate hairs. *Corolla* yellow, hairy outside; tube 14–15 mm long; limb 7–8 mm diameter, divided into rounded or oval, ± entire lobes. *Anthers* 1.2 mm long, inserted in the throat or near the middle of the corolla-tube. *Style* reaching into the throat or one third to half-way along the corolla-tube. *Capsule* with 3–4 seeds.

Distribution. Central S. Iran: Kerman Province, close to Kerman (Syrtsh, Kuh-i-Jupar); altitude 2100–3800 m.

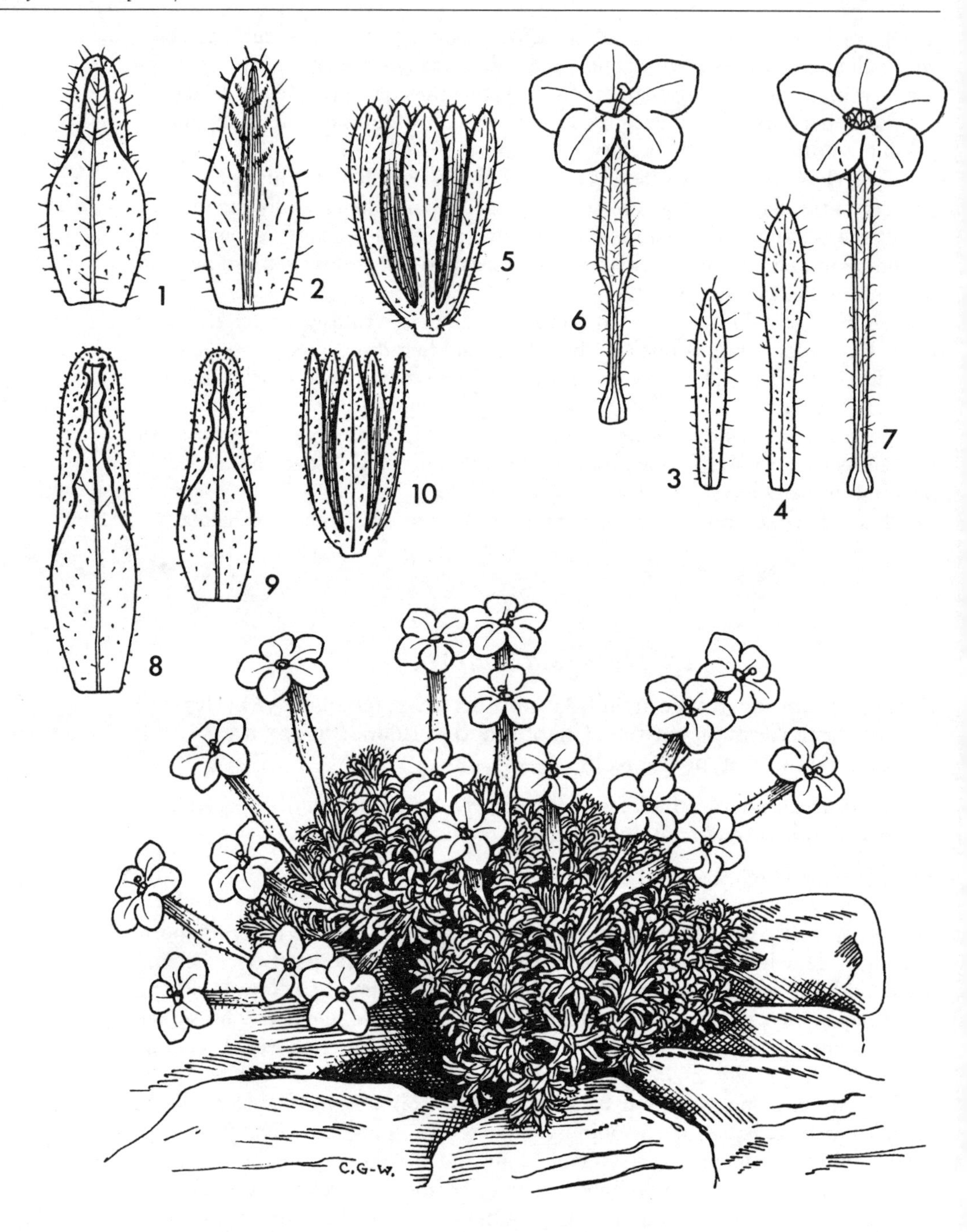

Dionysia rhaptodes, × 2. 1 leaf below, × 15; 2 leaf above, × 15; 3–4 bracts, × 9; 5 calyx, × 9; 6 pin-eyed corolla, × 4½; 7 thrum-eyed corolla, × 4½. *D. oreodoxa*. 8–9 leaves below, × 15; 10 calyx, × 9.

HABITAT. Growing in crevices on north-facing limestone cliffs. Flowering in April and May.

For general comments see under *D. oreodoxa*, p. 94.

Dionysia rhaptodes was first discovered and collected E of Kerman by Bunge in 1859. In 1892 Bornmüller also collected this species together with *D. oreodoxa* on Kuh-i-Jupar, S of Kerman. Bornmüller as well as describing *D. oreodoxa* also described, from his collections, *D. heterochroa* which he considered to be close to *D. rhaptodes*, differing in the glabrous leaves and in the corollas which turn greenish on drying. However, if one examines the dried specimens of *D. heterochroa* and *D. rhaptodes* a complete transition from almost glabrous to pubescent leaves, and yellowish to greenish corollas, can be observed. Indeed the type specimen of *D. heterochroa* has a few scattered hairs on the leaves – they are not glabrous as Bornmüller suggested. As a result it is not possible to maintain *D. heterochroa* and I follow Wendelbo (1961) in making it a synonym of *D. rhaptodes*.

Dr Giuseppi saw *D. rhaptodes* on Kuh-i-Jupar in 1932, reporting that – 'There were countless hundreds of rosettes of all sizes...'. This species is unfortunately not in cultivation.

14 *Dionysia oreodoxa*

Dionysia oreodoxa Bornm. in Bull. Herb. Boiss. 7: 68, t.2 fig. 1 (1899).
Type: Iran, Kerman, Kuh-i-Jupar, *Bornmüller* 3873b (holotype B).

Primula oreodoxa (Bornm.) *loc. cit.*

Primula kermanensis Bornm. in Bull. Herb. Boiss. Ser. 2,3: 592 (1903).

DESCRIPTION. *Plants* very like *D. rhaptodes* in overall appearance and general characteristics, but forming laxer cushions with larger leaf-rosetes. *Leaves* linear-oblong, 4–5.5 mm long, 0.8–1.5 mm wide, the margin crenate and revolute in the upper half, with 3–4 teeth on each side, sparsely covered in minute capitate-glands, but without hairs, often with yellowish farina beneath or in the leaf-axils; leaves of vegetative shoots often larger, to 10 mm long, without a revolute margin. *Bracts* linear-lanceolate, acute, glandular like the leaves. *Calyx* with linear-elliptical, acute lobes, covered in minute capitate glands. *Corolla* yellow, hairy outside; tube 13–15 mm long; limb 8–10 mm diameter.

DISTRIBUTION. Central S Iran: Kerman Province, close to Kerman (Kuh-i-Jupar, Kuh-i-Nasr and Kuh-Tajh-Ali); altitude 2300–3400 m.

HABITAT. Growing in crevices on north-facing limestone cliffs. Flowering in April and May.

Dionysia oreodoxa was first collected on 3 mountains in the Kerman area of central southern Iran, by Bornmüller in 1892. No further material was collected until 1966 when Jim and Janette Archibald ventured to this rather remote region.

With its toothed and revolute leaf-margins *D. oreodoxa* must be associated with subsection *Revolutae*. Bornmüller (1899) associated this species closely with

D. revoluta, but the two are quite different. *D. oreodoxa* has a deeply divided calyx and entire corolla-lobes and a few-seeded capsule. In contrast, *D. revoluta* has a rather shallowly divided calyx, deeply notched corolla-lobes and many seeds to each fruit-capsule.

Dionysia oreodoxa finds its closest ally in *D. rhaptodes*, which is also endemic to the Kerman region. Indeed these two species are the only ones in subsection *Revolutae* to have entire corolla-lobes. There is little doubt that *D. oreodoxa* and *D. rhaptodes* are very closely related. Wendelbo (1961) has suggested that – 'It would not be surprising, however, if a close study of natural populations were to reveal transitional leaf forms. The latter (*D. oreodoxa*) would then have to be reduced to a form or variety of *D. rhaptodes*.

Dionysia rhaptodes forms a dense plant with smaller, more closely packed leaves. The leaf-margin is entire or slightly wavy, in contrast to the markedly crenate margin of *D. oreodoxa*.

More recent material collected by Jim Archibald in 1966 shows both species to be rather more variable than formerly supposed. However, both taxa can still be clearly distinguished and must, at least for the present time, be kept as distinct, though closely allied, species.

Dionysia oreodoxa is not in cultivation, although Jim Archibald commented in 1969 that – '*D. oreodoxa* was not a proper seed collection – only some pulverised cushions and the debris shaken from out of the herbarium specimens ... nothing germinated in 1967 or 1968, but today (March 1969) I notice what is definitely a seedling Dionysia in the pan. I don't give much for its chances; however, bearing in mind its southerly desert habit'.

15 *Dionysia esfandiarii*

Dionysia esfandiarii Wendelbo in Bot. Not. 123: 302, fig. 2 m-n (1970).
Type: Iran, Abadeh, Bacanat, Kuh Khataban, June 1969, *Termé* 8128E (holotype GB; isotype TARI).

DESCRIPTION. *Plants* forming rather dense grey-green cushions (c. 7 cm diameter in the wild), efarinose. *Stems* covered in dead leaves and leaf remains, partly overlapping. *Leaves* linear-oblong to lanceolate, c.4 mm long, 0.6–8.8 mm wide, the apex obtuse, the margin entire, revolute in the upper half, covered in long articulate spreading hairs in the upper half, but with shorter erect hairs in the lower half. *Flowers* solitary, sessile. *Bracts* 2, linear, 2.5–4 mm long, acuminate, pubescent like the leaves. *Calyx* c. 4 mm long, divided for two-thirds of its length into linear-triangular lobes, pubescent. *Corolla* violet, pubescent outside; tube ? 13–14 mm long, limb ? 8–9 mm diameter, divided into obcordate, clearly emarginate lobes.

DISTRIBUTION. SW Iran: Fars; Abadeh, Bavanat, Kuh Khataban.

HABITAT. Unknown, probably limestone cliffs; altitude c 3100 m. Flowering in the wild probably in April.

Dionysia esfandiarii was discovered in 1969 by Fereydoun Termé, a botanist at the Department of Botany, Plant Pest and Diseases Research Institute, Tehran,

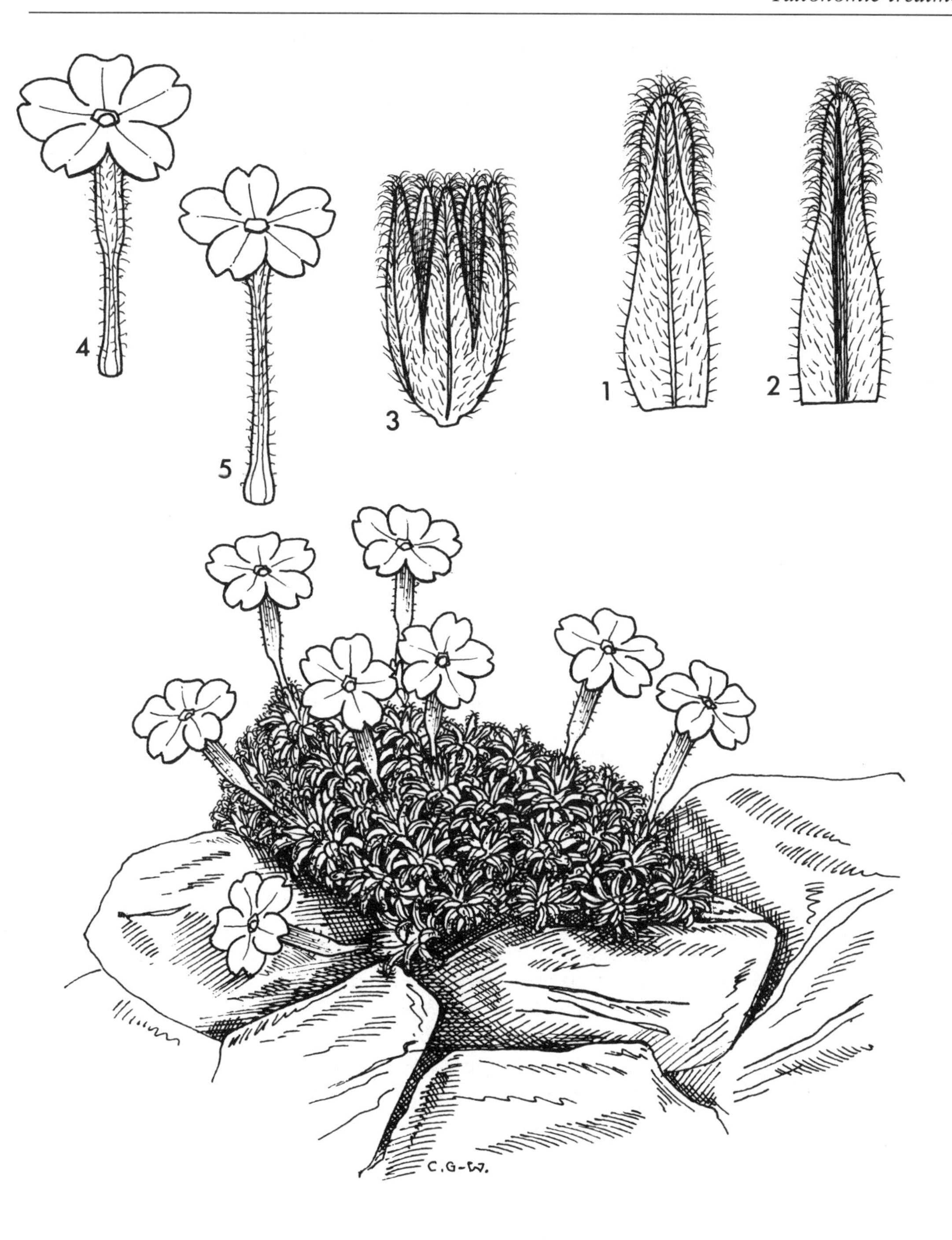

Dionysia esfandiarii, × 2. 1 leaf below, × 12; 2 leaf above, × 12; 3 calyx, × 9; 4 pin-eyed corolla, × 3; 5 thrum-eyed corolla, × 3 (corollas speculative).

and named in honour of Dr E. Esfandiari, the former head of the Herbarium, Iranian Ministry of Agriculture in Tehran.

The type specimen, the only collection of this species, is incomplete and in a young state and Wendelbo was unable to give a full description – the flowers for instance are not fully developed. However, *D. esfandiarii* is a very distinct species in subsection *Revolutae*. With its violet flowers and revolute leaf-margins it could be associated with *D. archibaldii*. However, *D. esfandiarii* is unique with the leaves in which the lower articulated hairs are short and upward directed, whereas those at the apex are longer and curving downwards – the same applies to the bracts and calyces. The leaf-margin is untoothed, like that of *D. rhaptodes*, but unlike it the corolla-lobes are clearly notched. *D. esfandiarii* therefore appears to hold a rather isolated position within subsection *Revolutae*.

Dionysia esfandiarii probably forms small dense cushions, silvery-grey and studded with small violet flowers. It would undoubtedly be a fine addition to cultivation but is likely to prove rather difficult to grow.

16 *Dionysia lurorum*

Dionysia lurorum Wendelbo in Notes Roy. Bot. Gard. Edinb. 38(1): 105, fig. 1 (1980). Type: Iran, Luristan (= Lorestan), 61 km on road from Aligodarz to Shoulabad (valley after the pass), 2400 m, 29 June 1977, *Runemark & Lazari* 26216 (holotype G; isotypes E, TARI).

DESCRIPTION. *Plants* form fairly lax cushions; branches thin, not becoming noticeably thick and woody with age. *Leaves* variable, those of flowering shoots borne in rather distant whorls, elliptic to suborbicular, 1.5–4 mm long, 1.5–2.5 mm wide, the apex acute to shortly apiculate, the margin entire or slightly crenulate; leaves of vegetative shoots larger, alternate, oblanceolate to elliptic or linear-elliptic, 5–9 mm long, 1.5–3.5 mm wide, the apex acute, the margin ± entire to remotely denticulate; all leaves with veins rather flabellate (at least in dried specimens), minutely glandular or both surfaces, with white woolly farina beneath, especially towards the base and in the axils. *Inflorescence* a 1–3-flowered umbel; flowers sessile, heterostylous. *Peduncle* 4–13 mm long, glandular-pubescent. *Bracts* foliaceous, ovate to lanceolate, 8–10 mm long, 3–4.5 mm wide, ± entire to remotely denticulate, glandular like the leaves. *Calyx* tubular-campanulate, 7–9 mm long, divided almost to the base into linear-lanceolate, acute, entire lobes, glandular, farinose towards the base. *Corolla* yellow, sparsely glandular-pubescent outside; tube 12–16 mm long; limb 9–11 mm diameter, divided into elliptic-obovate, entire lobes. *Anthers* c. 2 mm long, inserted just below the middle of the corolla-tube or in the throat. *Style* reaching just below the throat of the corolla-tube or about one-third the way along it. *Capsule* containing numerous seeds.

DISTRIBUTION. W Iran; Luristan, between Aligodarz and Shoulabad – not known elsewhere; altitude c. 2400 m.

HABITAT. Partially shaded moist limestone cliffs. Flowering during May and June.

Dionysia lurorum was discovered in 1977 by Runemark and Lazari. With its flat leaves, leaf-like bracts and short scape this interesting species clearly belongs to subsection *Caespitosae*. *D. lurorum* finds its closely ally in *D. caespitosa*, differing in the distinctive leaf glomerules, the calyx which is split almost to the base and the shorter corolla-tube. Furthermore, the style of the long-styled flower is never exserted and the presence of farina on the leaves is very distinctive. Indeed the short corolla-tube and farina distinguish this species from all other members of subsection *Caespitosae*.

According to Wendelbo (1980) the presence of farina, and the large number of seeds per fruit-capsule are considered primitive characters in *Dionysia*, this leading to the conclusion that *D. lurorum* is the most primitive member of subsection *Caespitosae*.

The species gets its name from the Luris, the inhabitants of the Iranian Province of Luristan. *D. lurorum* is not in cultivation.

17 *Dionysia caespitosa*

Dionysia caespitosa (Duby) Boiss., Diagn. Pl. Or. Nov. 1(7): 67 (1846).

Gregoria caespitosa Duby in DC. Prodr. 8: 46 (1844). Type: W Iran, Elvend Kuh, near Esfahan, *Aucher-Eloy* 2609 (lectotype K; isolectotype BM, P – chosen here).

Macrosyphonia caespitosa (Duby) Duby, Rev. Gen. Pl. 2: 400 (1891).

Dionysia peduncularis Bornm., Viert. Beitr. Kenntis Gatt. Dionysia in Bull. Herb. Boiss. 5: 261 (1905). Type: W Iran, Kohrud, April 1904, *Strauss* s.n. (holotype B).

DESCRIPTION. *Plants* forming rather dense green tufts or low cushions (to 20 cm diameter in the wild), efarinose. *Stems* well-branched, becoming woody below, above covered in spreading dead leaves and leaf-remains. *Leaves* oblong to obovate to oblanceolate, 3–6 mm long, 1.0–2.5 mm wide, obtuse to subacute, the margin subentire to toothed, with 1–8- small teeth towards the apex, covered on both side with stipitate glands and often with long articulate hairs mixed in; veins thin, reticulate, poorly defined; leaves of vegetative shoots larger, to 15 mm long and 3.5 mm wide, often stalked, subentire to ± toothed. *Flowers* borne in 2–3 flowered umbels, sometimes solitary, heterostylous. *Scape* 3–30 mm long, pubescent. *Bracts* generally 3–4, variable, ovate to obovate or elliptical, coarsely toothed to subentire, occasionally lobed, glandular like the leaves. *Calyx* broadly campanulate, 6–9 mm long, divided for about two-thirds of its length into ovate-lanceolate acute, entire lobes, covered in stipitate glands outside; tube 20–30 mm long; limb 11–12 mm diameter, divided into suborbicular entire lobes. *Anthers* 2 mm long, inserted in the throat or three-quarters the way up the corolla-tube. *Style* well exserted or reaching the middle of the corolla-tube. *Capsule* with 8–12 seeds.

subsp. **caespitosa**

Leaves subentire or with up to 3, often rather obscure teeth (Plate 18).

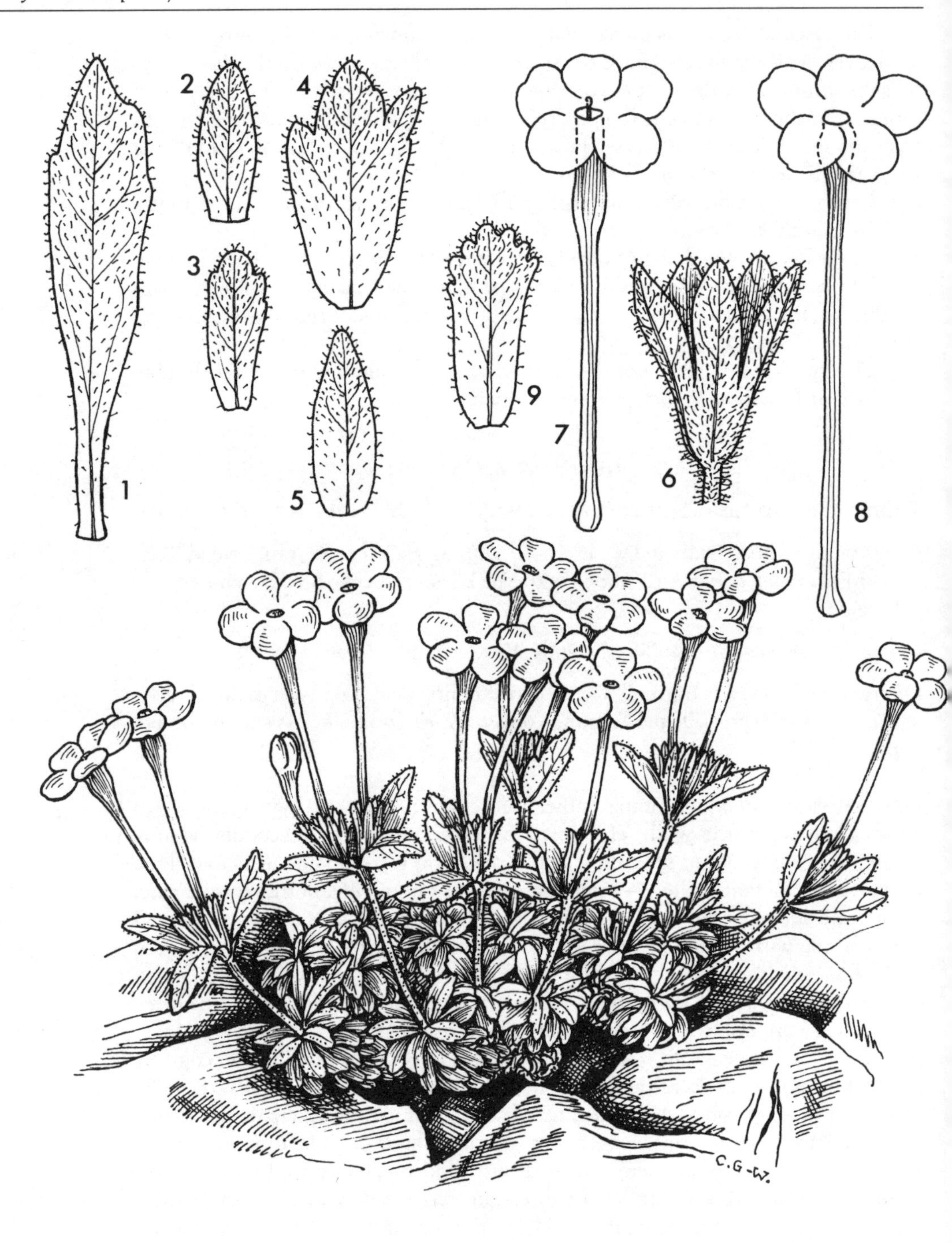

Dionysia caespitosa subsp. *caespitosa*, × 1½. 1 vegetative leaf, × 6; 2–3 leaves of flowering shoot, × 6; 4–5 bracts, × 6; 6 calyx, × 4½; 7 pin-eyed corolla, × 3; 8 thrum-eyed corolla, × 3. Subsp. *bolivarii*. 9 leaf, × 6.

DISTRIBUTION. W Iran; near to Esfahan (Elvend Kuh, Dumbe Kemer, Kuh Bärsukh, Kuh Gamser, Kohrud, Chehel Dokhtaran Kuh, Kuh-e-Darabshah); altitude 1700–3000 m.

HABITAT. Crevices in shaded or sunny limestone cliffs, often N- and NW-facing, occasionally on screes or sloping limestone slabs; flowering in late April, May and early June.

subsp. **bolivarii** (Pau) Grey-Wilson, *comb. & stat. nov.*

Dionysia bolivarii Pau in Pau & Vicioso, Trab. Mus. Nac. Cienc. Nat. (bot.) 14: 27 (1918). Type: W Iran, Bazuft Valley, *Escalera* s.n. 1899 (holotype MA).

Leaves with regularly toothed margins, 4–8 small obtuse teeth in the upper third.

DISTRIBUTION. W. Iran; Bakhtiari, Bazuft Valley; altitude unknown.

HABITAT. Probably limestone cliffs; flowering in May and June.

Dionysia caespitosa was first collected by Aucher-Eloy in Iran, to the NW of Esfahan in 1835. Since then it has been collected on a number of mountains in the region. Some doubt has been expressed as to whether Aucher-Eloy collected the species on Elvend Kuh near Esfahan or the mountain of the same name near Hamadan. However, this species has been collected on the mountain near Esfahan, but never near Hamadan, so there seems little doubt that the type locality must be Elvend Kuh near Esfahan.

Duby (1844) first described this species as *Gregoria caespitosa* on account of its scapose inflorescence and large leafy bracts, but later in the same year he made a new genus, *Macrosyphonia*, for it. To add to this confusion Clay (1937) proposed that this species, together with *Dionysia bornmuelleri, D. hissarica* and *D. lacei* be placed in a genus of their own, *Dionysiopsis*. Fortunately this was never done. There is little doubt from anatomical, cytological and palynological evidence that this is a species of *Dionysia*.

The scapose inflorescence and leafy bracts make *D. caespitosa* one of the more primitive members of subsection *Caespitosae*, along with the recently described *D. lurorum*. The species is quite variable in leaf and bract characteristics – they may be narrow to broad, subentire to few-toothed.

Dionysia bolivarii, collected first by Manuel Martinez De La Escalera in the Bazuft Valley (Bakhtiari Province) in 1899 differs only in the more distinctly and evenly toothed leaf-margins. As it is geographically isolated to the south of *D. caespitosa* I have treated it as a subspecies of that species.

Dionysia caespitosa is not in cultivation today; however, it was for a short while a few years ago. Seed was introduced by Paul Furse in 1962 and Jim Archibald in 1966. Only the latter collections germinated. Plants were successfully raised and the first flowered in 1969. However, it quickly proved one of the most temperamental species in cultivation, particularly as regards watering. Jack Elliott wrote at the time – 'I have one very small plant which I potted but it only just survived the operation'. The late Peter Edwards also commented – 'This is one of the more difficult species in every way. It forms a loose cushion ... I lost my original plant following repotting and I would therefore say that it would be unwise for anyone to follow my example, although I did have the good sense to root a cutting before I attempted it. Care with watering at all times, just keeping the compost moist in the growing period and fairly dry in winter'.

NOTE. In the *AGS Bulletin* 1965 (33: 360) the plant mentioned under the name of *D. caespitosa* is in fact *D. tapetodes*, the seed having been collected in Afghanistan.

18 *Dionysia gaubae*

Dionysia gaubae Bornm. in Fedde Repert. 41: 179 (1937). Type: Iran, Luristan, Khorramabad, Pole-Kalhor, April 1936, *Gauba* s.n. (holotype B).

DESCRIPTION. *Tufted perennial* forming small cushions, efarinose; branches woody and bare below, clothed with spreading and overlapping or reflexed marcescent leaves above. *Leaves* ovate to oblong or spathulate, the apex obtuse to subobtuse, the margin entire or with one or two blunt teeth in the upper half, covered on both surfaces with minute capitate glands. *Flowers* solitary, sessile, heterostylous – only the short-styled flower known. *Bracts* 2, subequal, linear-oblong to oblanceolate, 4.5–6 mm long, entire or with several blunt teeth, glandular. *Calyx* tubular-campanulate, c. 6.5 mm long, divided almost to the base into linear-oblong, entire lobes, glandular. *Corolla* yellow, glandular-pubescent outside; tube c. 24 mm long; limb 10–11 mm diameter, divided into obovate, entire lobes. *Anthers* c. 2 mm long, inserted in the throat of the corolla-tube. *Style* reaching the middle of the corolla-tube (short-styled flower). *Capsule* with c. 3 seeds.

DISTRIBUTION. W. Iran, Luristan, Pole-Kalhor near Khorramabad – not known elsewhere; altitude c. 1000–1160 m.

HABITAT. Growing on steep limestone cliffs. Flowering in March and April.

Dionysia gaubae appears to be a very rare and possibly relict species which has only ever been collected twice in the wild. The species was discovered in April 1936 by Erwin Gauba, an Austrian agriculturist who was Professor at the Agricultural College of Karaj in N. Iran before the Second World War. Gauba found the plant not far from the city of Khorramabad in the same vicinity as he found the rare *Primula gaubaeana* Pax. The specimen of *D. gaubae* which is in the Berlin Herbarium is small and is probably all that Bornmüller based his description on.

The only other collection of *D. gaubae* was made by Jim Archibald in April 1966. At the time he wrote that this species was found on 'the left bank of the Kashgan Rud above the Pol-i-Khaller, 60 km west of Khorramabad – *D. gaubae* was not in flower – in spite of exhausting searching we only found a few cushions which were quite impossible to photograph. It is little more than a relict form of *D. caespitosa* stranded at a low altitude and certainly dying out. It may occur in better health in some other gorges of the Kasghan Rud ... No seed of *D. gaubae*

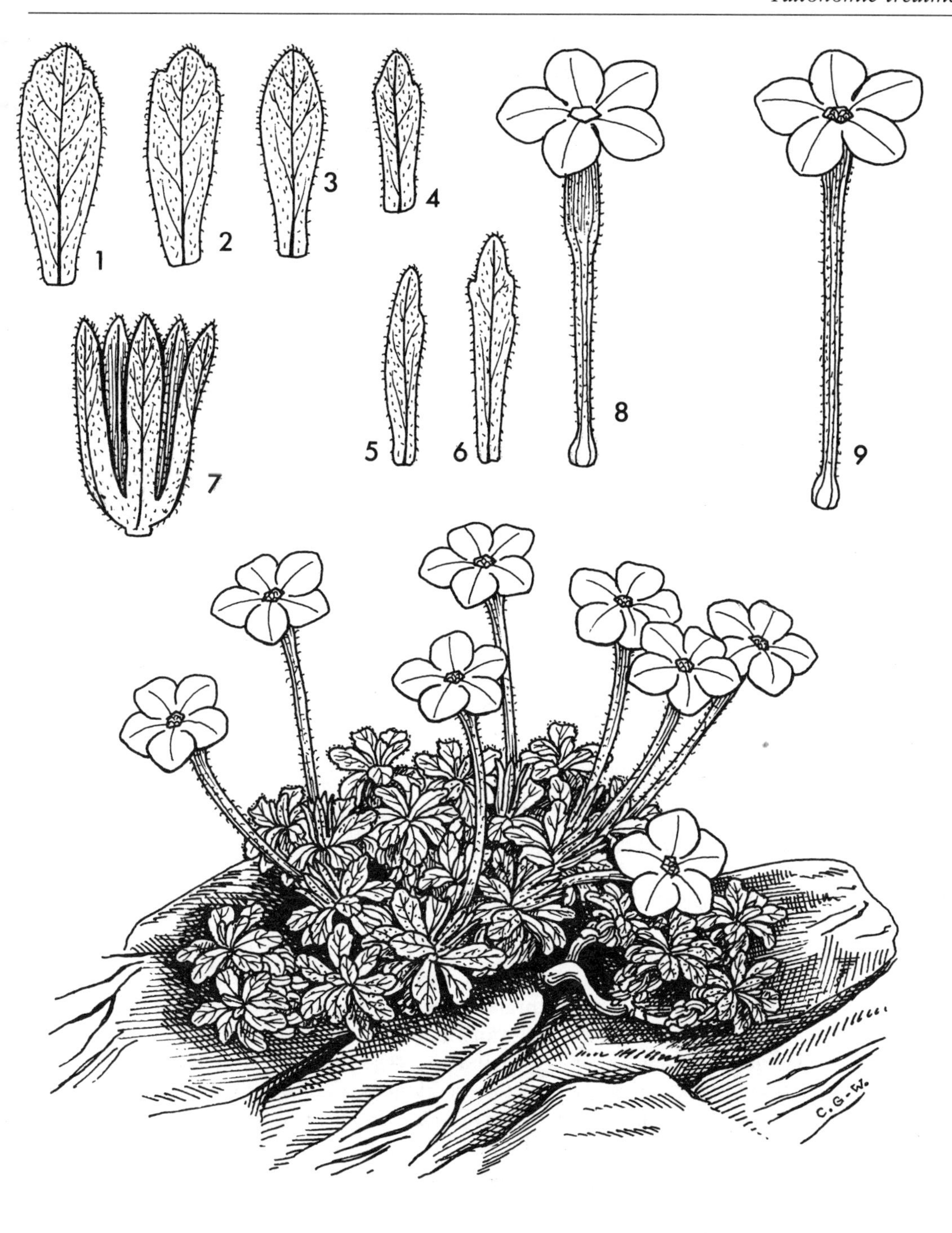

Dionysia gaubae, × 2. 1–4 leaves, × 6; 5–6 bracts, × 6; 7 calyx, × 6; 8 pin-eyed corolla, × 3; 9 thrum-eyed corolla, × 3.

was collected, although I visited the site again in the summer (when we collected seed of *Primula gaubaeana*). We were lucky to get the miserable herbarium specimens without loss of life or limb and found only a handful of plants – I counted 9 reasonably healthy ones'.

Dionysia gaubae, then, is a little known species. Wendelbo has suggested that it may be only 'a reduced *D. caespitosa*', however, as far as our present knowledge goes the two are clearly separable on geographical and morphological characteristics. *Dionysia gaubae* differs in its solitary, scapeless flowers and relatively larger leaves. It may be that in the Zagros Mountains other localities for both *D. gaubae* and *D. caespitosa* may exist and these may help to reveal the full range of variation within these species. For the present it seems best to regard them as distinct.

Dionysia gaubae has never been in cultivation.

19 *Dionysia diapensiifolia*

Dionysia diapensiifolia Boiss. Diagn. Pl. Orient. Nov. 1(7): 65 (1846). Type: Iran, Fars, near Persepolis, *Kotschy* 236 (holotype G; isotypes BG, BM, K, O, W).

Primula diapensiifolia (Boiss.) Kuntze, Rev. Gen. Plant. 2: 400 (1981).

Dionysia diapensiifolia Knuth in Pax & Knuth, Pflanzenreich 4 (237): 168 (1905).

Dionysia diapensiifolia Melchior in Mitt. Thür. Bot. Ver. N.F. 50: t.3, fig. 13, t.4 (1943).

Dionysia drabaefolia Bunge in Bull. Ac. Imp. Sci. St-Petersb. 16: 558 (1871). Type: Iran, Fars, near Persepolis, *Kotschy* s.n. (holotype ?LE; isotype P).

Dionysia drabifolia Knuth in Pax & Knuth, Pflanzenreich 4(237): 167 (1905).

Dionysia straussii Balls on Gard. Chron. Ser. 3, 95: 79, fig. 35 (1934), *non* Bornmüller.

DESCRIPTION. *Suffrutescent plant* forming large cushions up to 1 m across, though often less, efarinose; branches becoming woody below, densely covered in spreading marcescent leaves, forming closely packed columns. *Leaves* variable, obovate to oblong, elliptic or spathulate, 5–10 mm long, 2–3 mm wide, the apex obtuse, the margin entire or with several blunt, shallow or coarse, teeth on either side, especially in the upper half, densely covered on both surfaces with articulated glandular-hairs. *Flowers* solitary or occasionally 2 together, subsessile or occasionally with a short scape to 4 mm long, heterostylous. *Bracts* 2, rather unequal, linear to oblanceolate, 5.5–10 mm long, entire or slightly toothed, glandular-pubescent. *Calyx* tubular-campanulate, 5–10 mm long, divided for three-quarters of its length into oblong to linear, usually entire lobes, glandular-pubescent. *Corolla* yellow, glandular-pubescent outside; tube 16–30 mm long; limb 10–12 mm diameter, divided into ovate or suborbicular, entire or very

46 N. slopes of the Koh-i-Alborz, N. Afghanistan, on whose cliffs both *D. freitagii* and *D. hedgei* were discovered. Photo. C. Grey-Wilson.

47 *Dionysia hedgei* growing beneath the limestone overhangs, Koh-i-Alborz, N. Afghanistan. Photo. C. Grey-Wilson.

48 *Dionysia hedgei* growing on sloping limestone slabs, Koh-i-Alborz, N. Afghanistan. Photo. C. Grey-Wilson.

49 *Dionysia freitagii*, a 10 year old plant.
Photo. E. G. Watson.

50 *Dionysia freitagii*. Photo. S. Taylor.

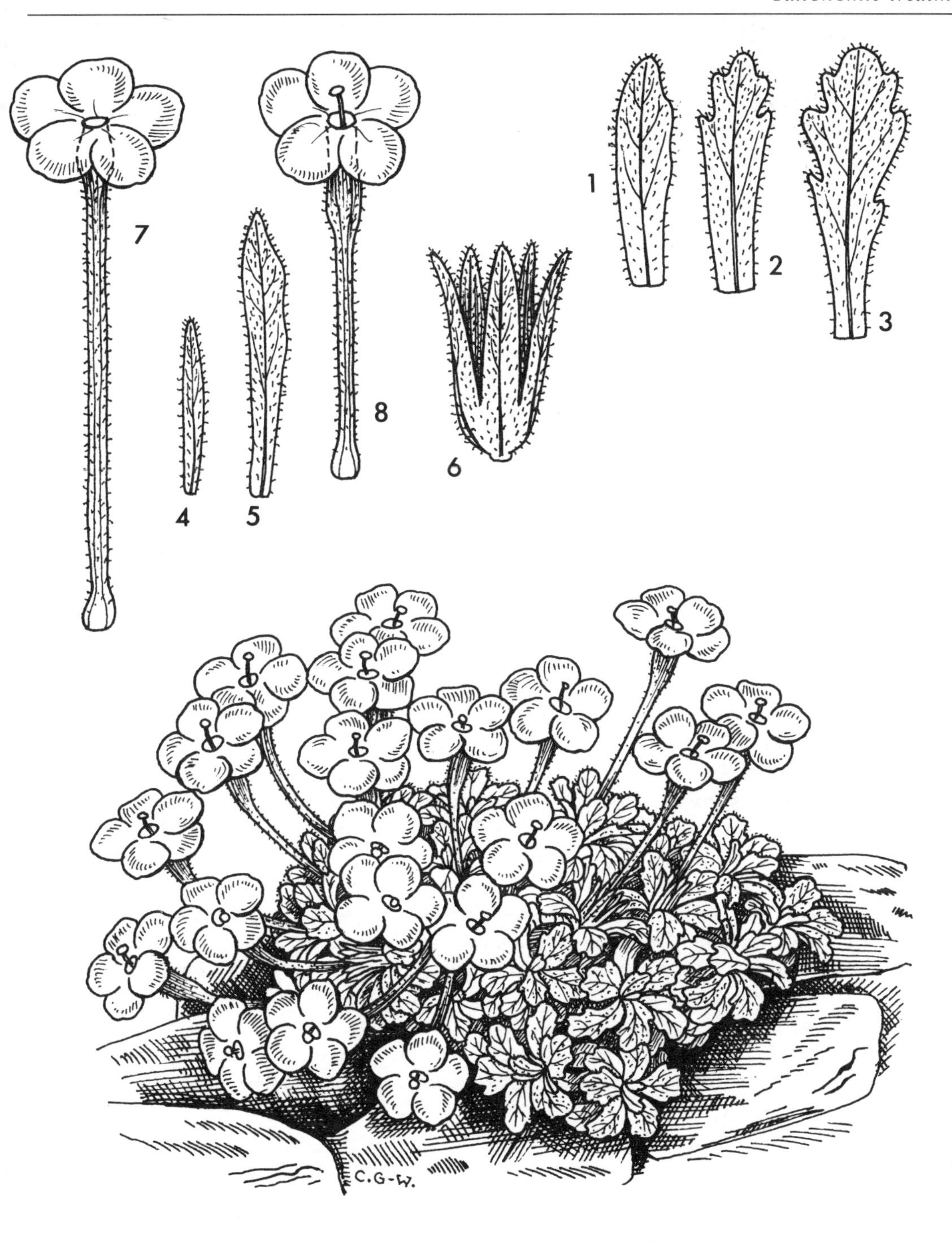

Dionysia diapensiifolia, × 2. 1–3 leaves, × 4½; 4–5 bracts, × 4½; 6 calyx, × 4½; 7 thrum-eyed corolla, × 3; 8 pin-eyed corolla, × 3.

slightly emarginate lobes. *Anthers* c. 2 mm long, inserted in the throat of the corolla-tube or just below it. *Style* reaching the middle of the corolla-tube or exserted. *Capsule* with 5–6 seeds usually (Plates 1, 19–20).

DISTRIBUTION. SW Iran, Fars Province – near Persepolis, between Persepolis and Daulatabad, Nakshi Rustam, Kuh Ayub, Kuh Bungi, Kuh-i-Sabzpuchon, Kuh-i-Dinar; altitude 1600–2300 m.

HABITAT. Growing on shaded and semi-shaded limestone cliffs, generally of north or north-east aspect. Flowering during March and April.

Dionysia diapensiifolia finds its closest allies in the members of subsection *Caespitosae*, particularly *D. odora*, from which it differs primarily in its far larger cushions in which the leaves are spreading rather than closely imbricating and in the generally larger flowers. In *D. odora* the leaves are generally only toothed at the apex, whereas in *D. diapensiifolia* they are toothed in the upper half or two-thirds.

In many respects *D. diapensiifolia* bridges the gap between the scapose *D. caespitosa* and *D. lurorum*, and the scapeless *D. odora* and *D. termeana*. In *D. diapensiifolia* there is often a very short scape, 1–4 mm long.

The species was first collected by Aucher-Eloy on Elvend Kuh, near Esfahan, in 1835.

Dionysia diapensiifolia has a very long history. In 1605 Clusius pictured two species of *Dionysia* in his *Exoticorum libri decem* as *Amomum quorundum* and *Pro Amonide data*. These specimens had been sent to Walerandus Donraeus in Lyon from Ormuz, at that time an important port situated near Bandar Abbas on the Persian Gulf. The upper drawing of Clusius is undoubtedly *D. diapensiifolia* and Wendelbo (1961) has suggested that the lower represents *D. rhaptodes*. The former is common around Shiraz and the latter in the vicinity of Kerman, both important cities in S. Iran and both with good connections with the harbour of Ormuz during the seventeenth century.

The identity of a drug sold in Bombay for some centuries under the name *Amamon Amooman* or *Hamama* was unknown until Holmes ('The botanical identification of Hamama') in *Pharm. Jour. Trans.* 252 (1887) showed that it was a species of *Dionysia*. A portion of this drug purchased from a Bombay market is undoubtedly *D. diapensiifolia*, so that the plant must have been exported along the coast from Persia to India.

Interestingly, on Kotschy's label of the type specimen is written the words 'Tscha mama', a name close to *Hamama* (one of the names under which the drug was sold).

Dionysia diapensiifolia forms the largest cushions of any species of *Dionysia* in the wild, sometimes reaching 80–90 cm diameter, occasionally more. In cultivation, however, a plant 15 cm diameter (after about 5 years) is exceptional.

Seed has been introduced on a number of occasions in the 1960s and 1970s – most notably by Paul Furse, Jim Archibald and later by Grey-Wilson and Hewer and Anne Pickering (1977). At one time there were quite a few plants in cultivation and in the early 1970s the species made quite frequent appearances at Alpine Garden Society shows. However, since then the species has suffered a general decline and it is now considered difficult in cultivation. Its close ally, *D.*

caespitosa is even more difficult to grow, see p. 99.

Dionysia diapensiifolia makes mounds of dense foliage. The leaves are rather spreading and relatively large and this may be its chief failing for botrytis all too readily attacks the foliage and stems during damp dull weather during the autumn and winter. The soft yellow, long-tubed flowers are very elegant, reminiscent of winter jasmine, though not particularly freely produced.

Despite the fact that both pin- and thrum-eyed plants exist in cultivation no seed has ever been produced (even with hand pollination) and new plants must be raised from cuttings. Cuttings are not easy and a low success rate, 20–30%, must be expected; sometimes no cuttings in a batch will root. It is often difficult to find decent cuttings and Brian Burrow reports that – 'Cuttings tend to rot rather than root'.

Eric Watson says that the species is – 'Slow-growing – very difficult to keep and must be kept going by cuttings or it will be lost to cultivation'.

Dionysia diapensiifolia then has a tenuous existence in cultivation and it is solely the skill and dedication of a handful of growers that has kept it 'going' for the past twenty years.

20 *Dionysia odora*

Dionysia odora Fenzl., Flora 26: 390 (1843). Type: Iraq, Kurdisan, Mt Gara, *Kotschy* 386 (holotype W; isotypes BM, B, G, K).

Primula odora (Fenzl) Kuntze, Revisio Gen. Plant. 2: 400 (1891).

Gregoria aucheri Duby in DC. Prodr. 8: 46 (1844). Type: Iran?, Nal Kuh, *Aucher Eloy* 2832 (holotype G. isotypes BM, K).

Dionysia aucheri (Duby) Boiss., Fl. Orient. 4: 19 (1879).

Dionysia straussii Bornm. et Hausskn. in Bull. Herb. Boiss. ser. 2, 3: 591, t.6 fig. 1 (1903). Type: Iran, Kuh Gerru, near Burudjird, June 1902, *Th. Strauss*.

Dionysia odora subsp. *straussii* (Bornm. et Hausskn.) Bornm. in Beih. Bot. Centralbl. 28(2): 462 (1911).

Dionysia odora var. *straussii* (Bornm. et Hausskn.) Bornm. loc. cit. 33(2): 167 (1915).

Dionysia odora var. *integrifolia* Bornm. loc. cit. 33(2): 167 (1915).

Dionysia sintenisii Stapf ex Bornm. in Bull. Herb. Boiss. ser. 2,3: 592, t.6 fig. 3 (1903). Type: Turkey, Mardin, Bakahri, *Sintenis* 1282 (holotype B; isotypes G, K).

Aretia longiflora Fischer *nom. nud.*

DESCRIPTION. *Suffrutescent plant* forming rounded dense cushions to 50 cm across, generally less, efarinose; branches becoming woody below, covered in spreading to ± imbricate marcescent leaves, the lower often in distinct whorls. *Leaves* very variable, oblong to obovate, oval or spathulate, 3.5–4.5 mm long, 1–2 mm wide, the apex obtuse to subacute, the margin with 3–7 crenate to dentate

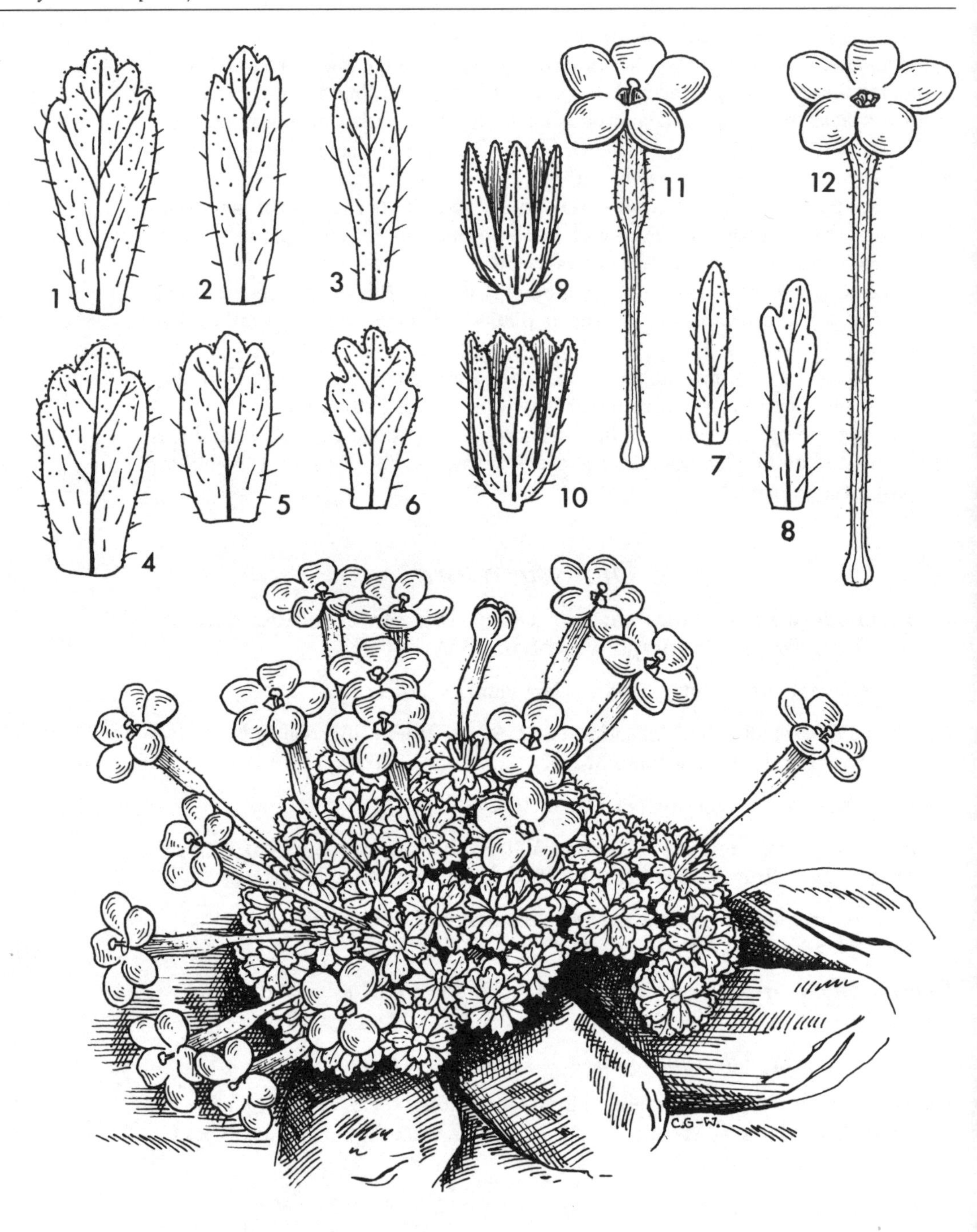

Dionysia odora, × 2. 1–6 leaves, × 9; 7–8 bracts, × 9; 9–10 calyces, × 4½; 11 pin-eyed corolla, × 3; 12 thrum-eyed corolla, × 3.

teeth towards the top, occasionally entire, covered on both surface with minute capitate glands and articulated hairs, rarely without hairs. *Flowers* solitary, subsessile, heterostylous. *Bract* solitary, linear-oblong to linear-oblanceolate, entire or with several teeth, glandular and pubescent like the leaves. *Calyx* tubular-campanulate, 3.8–4.5 mm long, divided for c. three-quarters of its length into linear-lanceolate to linear-oblong, entire or slightly toothed, lobes, glandular and pubescent. *Corolla* yellow, pubescent outside; tube 18–26 mm long; limb 7–10 mm diameter, divided into obovate or suborbicular, entire lobes. *Anthers* c. 1.5 mm long, inserted in the throat or two-thirds the way along the corolla-tube. *Style* reaching half-way along the corolla-tube or exserted. *Capsule* with c. 3 seeds.

DISTRIBUTION. Turkey, Mardin Province – Bakakri. Iraq, Kurdistan – NW and S of Sulaimanya, Mt Gara. W Iran, numerous mountains around Hamadan and Kermanshah and Khorramabad – Kasr-i-Shirin, Kuh-i-Kinischt, Kuh Milleh Michan, Kuh Parrau, Kuh-i-Retschab, Noa Kuh near Kerind, Safed Kuh; altitude 1100–2400 m.

HABITAT. Shaded and semi-shade rocks and cliffs, generally of limestone. Flowering during March and April.

Dionysia odora has the widest distribution of any member of section *Anacamptophyllum*, occuring as it does from SE Turkey to NW Iraq and W Iran east to Khorramabad and Hamadan. As a consequence, as might be expected, it is a very variable species which in part explains its rather extensive synonymy.

This species is particularly variable in leaf-shape and in the margin which ranges from entire to markedly toothed. There is a similar range of variation in the bracts and calyx-lobes and in the degree and density of the indumentum on all these organs. The size of corolla also varies from plant to plant. This amount of variation may appear to be useful for dividing the taxon and indeed several distinct entities have been previously recognised – for instance *D. straussii* Bornm. & Hausskn. described in 1903 was based on a plant with narrow leaves toothed all along the margin, not just at the apex. Another, collected from the type locality of *D. odora* and with entire leaves (*Poore* 438) may equate with *D. straussii* var. *glabrata*.

However, there is no significant pattern to any of the variations, indeed plants can be found in the same population differing markedly in corolla-size, leaf-indentation and the amount of pubescence. Wendelbo (1961) pointed out that – '... even in a single tuft some branches may have strikingly different leaves from the rest of the specimen'. As a result I am following Wendelbo in not recognising any formal subdivisions of *D. odora*.

Dionysia sintenisii Stapf pictured by Bornmüller (Bull. Herb. Boiss. ser. 2–3, Pl. Vl, fig. 3) was collected apparently near Mardin in SE Turkey. However, in 1957 Ian Hedge and Peter Davis searched there extensively for *D. odora* without success. Wendelbo (1961) comments that – '... they felt rather sceptical about the Sintenis gathering. In any case, it cannot be common at this north-westerly outpost of the genus'.

Dionysia odora finds its closest allies in *D. diapensiifolia* and *D. gaubae*. Both species have solitary escapose flowers, differing from other members of subsection *Caespitosae* in the lack of a scape. *D. odora* differs from both these species in the smaller, closely imbricating, leaves and rather smaller flowers.

It is perhaps very surprising that this attractive species is not in cultivation, but the reason is that the region in which it occurs has not been greatly explored botanically in recent years, partly due to political unrest and in Iran at least, due to the fact that most plant-collectors have concentrated on the mountains further to the south and east.

The first attempt to get *D. odora* into cultivation was made by the late Oleg Polunin in 1959 when he came across it on top of Piramagrun near Sulaimanya in N Iraq. However, this attempt failed and to quote Oleg Polunin's own words (*Bull. Alp. Gard. Soc.* 28: 229, 1960) – '... *D. odora* cuttings, like *D. bornmuelleri*, expired in the heat of Baghdad, and in the end I failed to bring back this treasure alive'.

21 *Dionysia termeana*

Dionysia termeana Wendelbo in Bot. Not. 123: 306 (1970), fig. 2g-1.
Type: Iran, Sisakht, Kuh-i-Daena to Gadaneh-Bidjan, June 1969, *Termé* 8131E (holotype GB; isotype Iran).

DESCRIPTION. *Plants* forming lax tufts or cushions, efarinose. *Stems* becoming woody below eventually, with the dead leaves and leaf-remains in distinct whorls, closely overlapping within each whorl. *Leaves* variable obovate-spathulate to linear-lanceolate or linear-oblong, 6–11 mm long, 2–5 mm wide, the outer in each leaf-rosette generally larger and broader than the inner, the apex obtuse to subacute, the margin entire or with up to 5–6 crenations on each side, glandular-pubescent on both surfaces; veins on all leaves rather distinct. *Flowers* solitary or occasionally 2–4 to a leaf-rosette, sessile, heterostylous. *Bracts* usually 2, linear-elliptical to oblanceolate, 2.5–7 mm long, glandular-hairy like the leaves. *Calyx* narrow-campanulate, 5–6 mm long, divided almost to the base into linear lobes, glandular-pubescent. *Corolla* yellow, glandular-pubescent outside; tube 15–17 mm long; limb 10–12 mm diameter, the lobes obcordate, emarginate. *Anthers* 1.5–2 mm long, reaching one third way along the tube or inserted below the throat. *Style* reaching one third or two thirds the way along the corolla-tube. *Capsule* containing c. 8 seeds (Plate 21–2).

DISTRIBUTION. SW Iran; Fars Province; Zagros Mountains, Sisakht, Kuh-i-Daena (= Dinar) to Gadaneh-Bidjan.

HABITAT. Shaded N and NW facing limestone cliffs, in narrow crevices and beneath overhangs; altitude 2700–3400 m. Flowering in the wild in April and May.

Dionysia termeana was first discovered and collected in June 1969 by Fereydoun Termé, an Iranian botanist working at the Ministry of Agriculture in Tehran. The species was named in his honour by Per Wendelbo.

This interesting plant, with its large lax cushions, generally toothed leaves and relatively large yellow flowers, has a distinctive look. The species appears, as far as it is known, to be confined to the Southern Zagros Mountains of Iran, in the

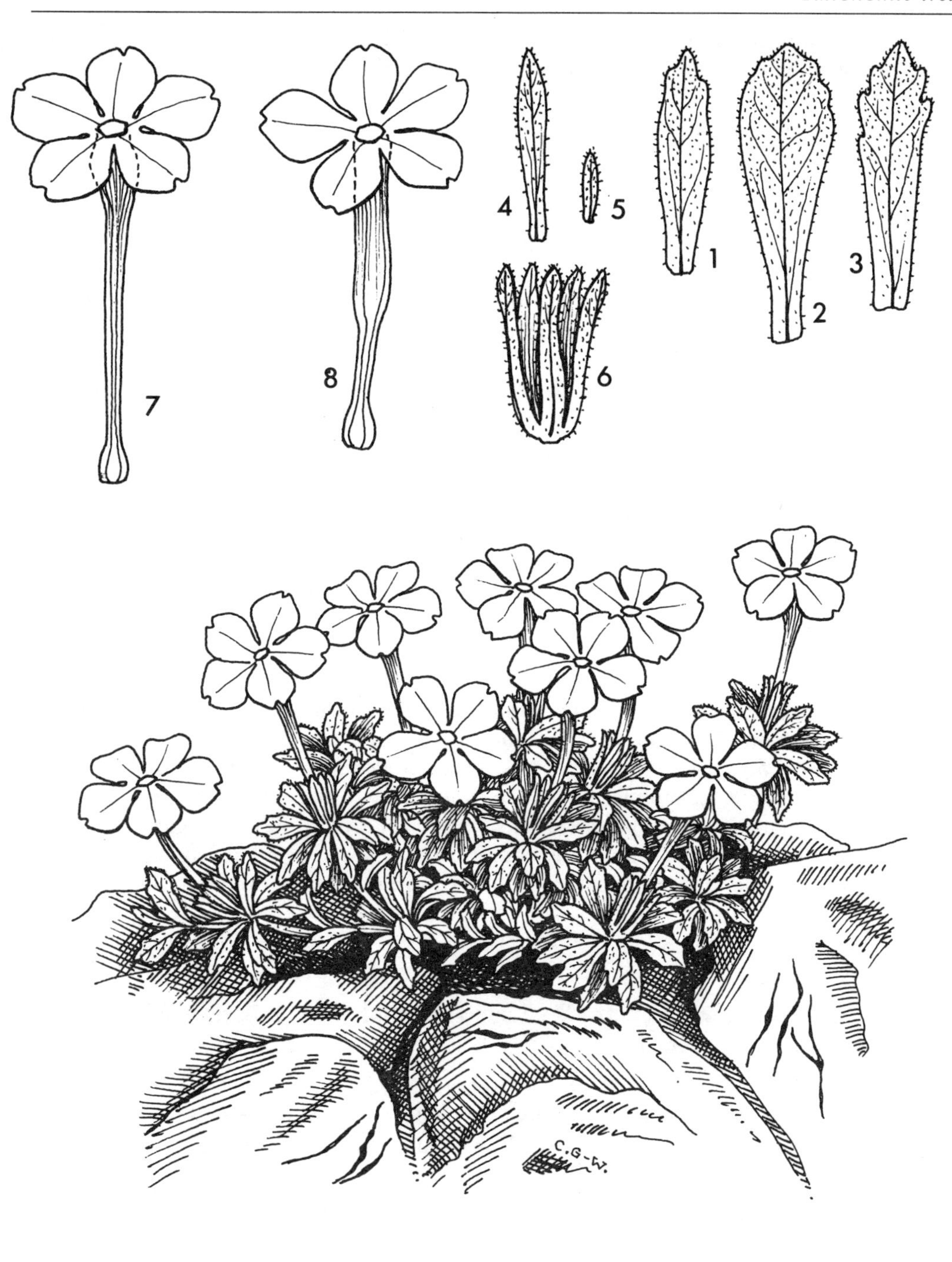

Dionysia termeana, × 2. 1–3 leaves, × 4½; 4–5 bracts, × 4½; 6 calyx, × 4½; 7 thrum-eyed corolla, × 3; 8 pin-eyed corolla, × 3.

vicinity of Kuh-i-Dinar near the town of Sisakht. Plants form small colonies in the wild on limestone conglomerate cliffs, especially favouring underhangs and cave entrances.

Dionysia termeana was subsequently re-collected by Davis & Bokhari in early April 1974 and then by Tom Hewer in early May 1973 when plants were seen in flower and photographed. All these collection show that *D. termeana* is quite variable, especially as regards leaf size, shape and the amount of toothing along the margins (they are sometimes practically entire). Leaf-rosettes generally bear only a solitary scapeless flower, but occasionally (as in *Hewer* 1987), two, three or even four flowers may be borne by a single leaf-rosette.

Clearly with its flat, sometimes somewhat involute leaves *D. termeana* belongs to section *Dionysia*, but there seems to be some confusion as to which subsection this species truly belongs to. Wendelbo (1970) in his original description of the species places it in subsection *Caespitosae*, alongside such species as *D. caespitosa* and *D. diapensiifolia*. However, in another publication (1976) Wendelbo transfers the species to subsection *Bryomorphae* on account of its lack of leaf sclereids, which all the other members of subsection *Caespitosae* bear. In general characteristics *D. termeana* looks far closer to the *Caespitosae* than to the *Bryomorphae*. The relatively large, toothed leaves, covered in capitate stipitate glands, are a characteristic of all members of the *Caespitosae*. The obcordate corolla lobes are more unusual. However, both *D. caespitosa* and *D. diapensiifolia* can have weakly notched or entire corolla-lobes and this character therefore is not a determining factor. The absence of leaf-sclereids clearly is a feature that would place *D. termeana* closer to the *Bryomorphae*, but in that subsection *D. curviflora* is the only species known to possess leaf-sclereids and there is no suggestion that for this character alone it should be transferred to the *Caespitosae*. In summarizing the position of *D. termeana* it is fair to say that it does not fit neatly into either subsection. The overall morphological characters would seem to place it in subsection *Caespitosae* and here I am content to leave it until further and more detailed evidence to the contrary comes to light.

22 *Dionysia sawyeri*

Dionysia sawyeri (Watt) Wendelbo in Årbok Univ. I. Bergen, Mat.-Nat. Ser. 3: 64 (1961).

Primula sawyeri Watt, Report of the botanical collections made in SW Persia by Major H. A. Sawyer, appendix (1891); Bornm. in Mitt. Thür. Bot. Ver. N.F. 47: 137 (1941). Type: W Iran, Kar Kanun, Kuhrang, *Sawyer* s.n. (not found, probably lost).

Dionysia bachtiarica Bornm. et Alex. in Bull. Herb. Boiss. 4: 515, t.2 fig. 3 (1905). Type: W Iran, Kellar, *Alexeenko* 2722 (holotype B).

Primula bachtiarica (Bornm. et Alex.) Bornm. *loc. cit.*

Description. *Plants* forming dense greyish tufts, efarinose. *Stems* short-branched, covered in closely overlapping dead leaves and leaf-remains. *Leaves* oblong to oblong-ovate, c. 5 mm long, 1.3 mm wide, acute to subobtuse, margin

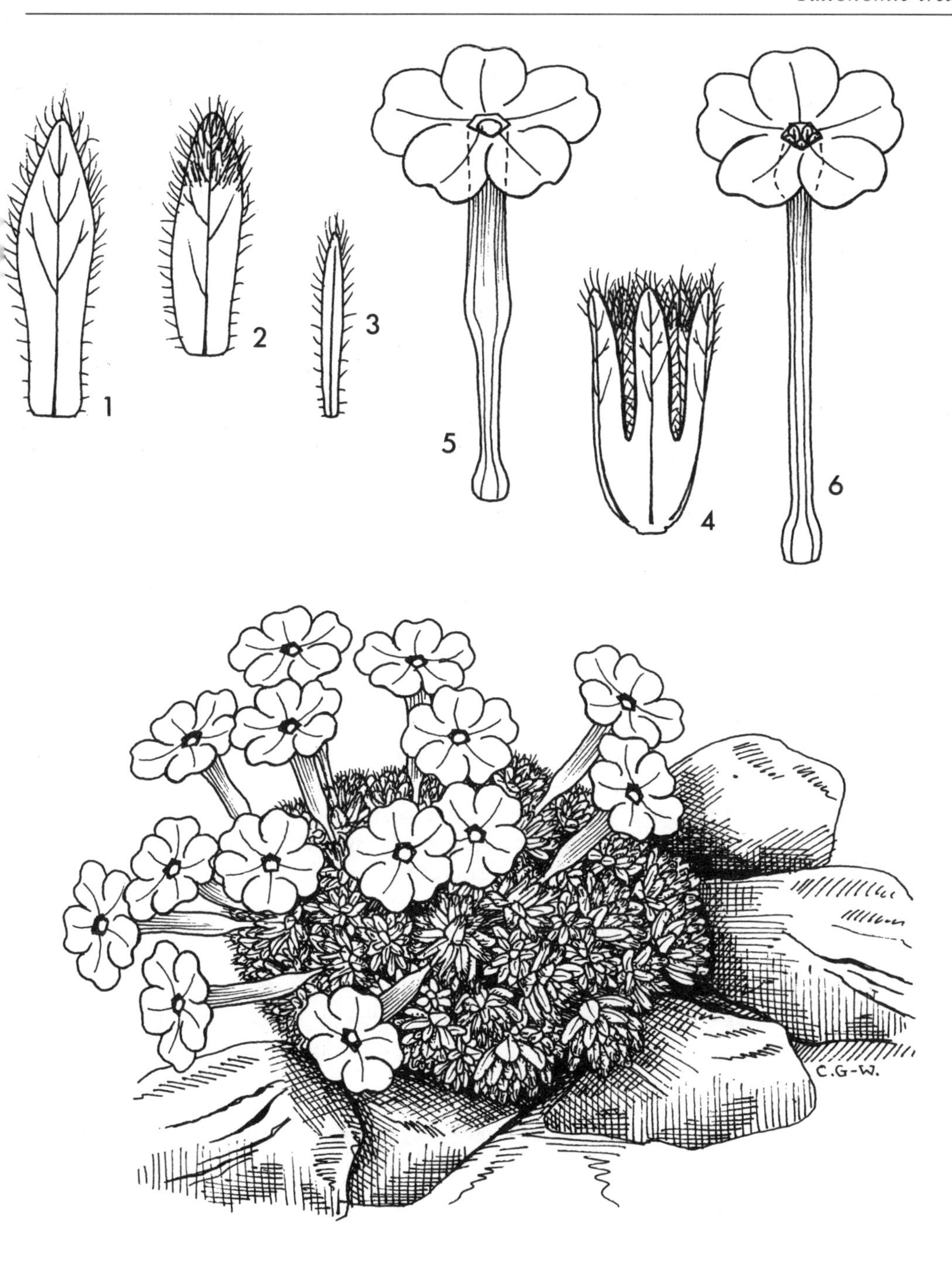

Dionysia sawyeri, × 2. 1 leaf below, × 9; 2 leaf above, × 9; 3 bracts, × 9; 4 calyx, × 4½; 5 pin-eyed corolla, × 4½; 6 thrum-eyed corolla, × 4½.

entire, with long articulate hairs along the margin, glabrous on both surfaces except sometimes for a few apical hairs; veins rather obscure, though midrib more prominent; leaves of vegetative shoots larger, to 15 mm long and 3 mm wide. *Flowers* solitary, subsessile, heterostylous (only the long-style flower form known). *Bract* 1, linear-subulate, c. 3 mm long, ciliate. *Calyx* tubular-campanulate, 4 mm long, divided for two-thirds of its length into narrow-oblong, ciliate lobes. *Corolla* violet-purple, glabrous; tube 12–13 mm long; limb 8–9 mm diameter, divided into obcordate, emarginate lobes. *Anthers* c. 1 mm long, inserted near the middle of the corolla-tube. *Style* reaching the throat of the corolla. *Capsule* with ?5–6 seeds.

DISTRIBUTIION. W. Iran; Bakhtiari Province, Kar Kanun and Kuhrang; altitude not known.

HABITAT and flowering time unknown.

Dionysia sawyeri was discovered in the Bakhtiari Mountains of Iran by Major H. A. Sawyer. G. Watt described the species in 1891 giving a brief description from the scanty material collected by Sawyer, and placing it in the genus *Primula* (as *Primula sawyeri* Watt). Bornmüller (1941) criticised Watts determination and therefore the species fell into obscurity, even being omitted from *Index Kewensis*.

Wendelbo (1961) had little doubt that *Primula sawyeri* was a species of *Dionysia* and transferred it to the genus. Unfortunately he was not able to trace the type specimen which should be at Edinburgh (E) or the Kew (K) herbarium and it is now thought to be lost.

At the same time, in 1904, Bornmüller and Alexeenko described a new species of *Dionysia*, *D. bachtiarica*, which as its name implies also hails from the Bakhtiari Mountains. Bornmüller did not (perhaps surprisingly!) connect the two plants, although there is little doubt that they are the same species.

Dionysia sawyeri has not been collected since Alexeenko saw it at the turn of the century. This is also perhaps surprising as the Bakhtiari Mountains have been explored by several expeditions in the past 30 years, although the mountains are extensive and often difficult of access.

Dionysia sawyeri has the largest leaves in subsection *Bryomorphae*. These have a neat edge of articulated hairs which give it a very distinctive look. The corolla was described as 'yellowish with a pale purple limb.'.

23 *Dionysia haussknechtii*

Dionysia haussknechtii Bornm. et Strauss in Bull. Herb. Boiss. 4: 514, t.2 fig 2 (1904). Type: Iran, Luristan, Shuturun Kuh. *Th. Strauss* s.n. 1903 (holotype B).

DESCRIPTION. *Plants* forming dense cushions, efarinose. *Stems* short-branched, becoming woody, contorted and leafless below, covered above with dead leaves, but these not overlapping to form columns. *Leaves* oblong, 2.5–4 mm long, 0.8–1.5 mm wide, obtuse, margin entire, covered above and beneath (especially in the upper half) in minute glandular-hairs, but with long articulate hairs forming a

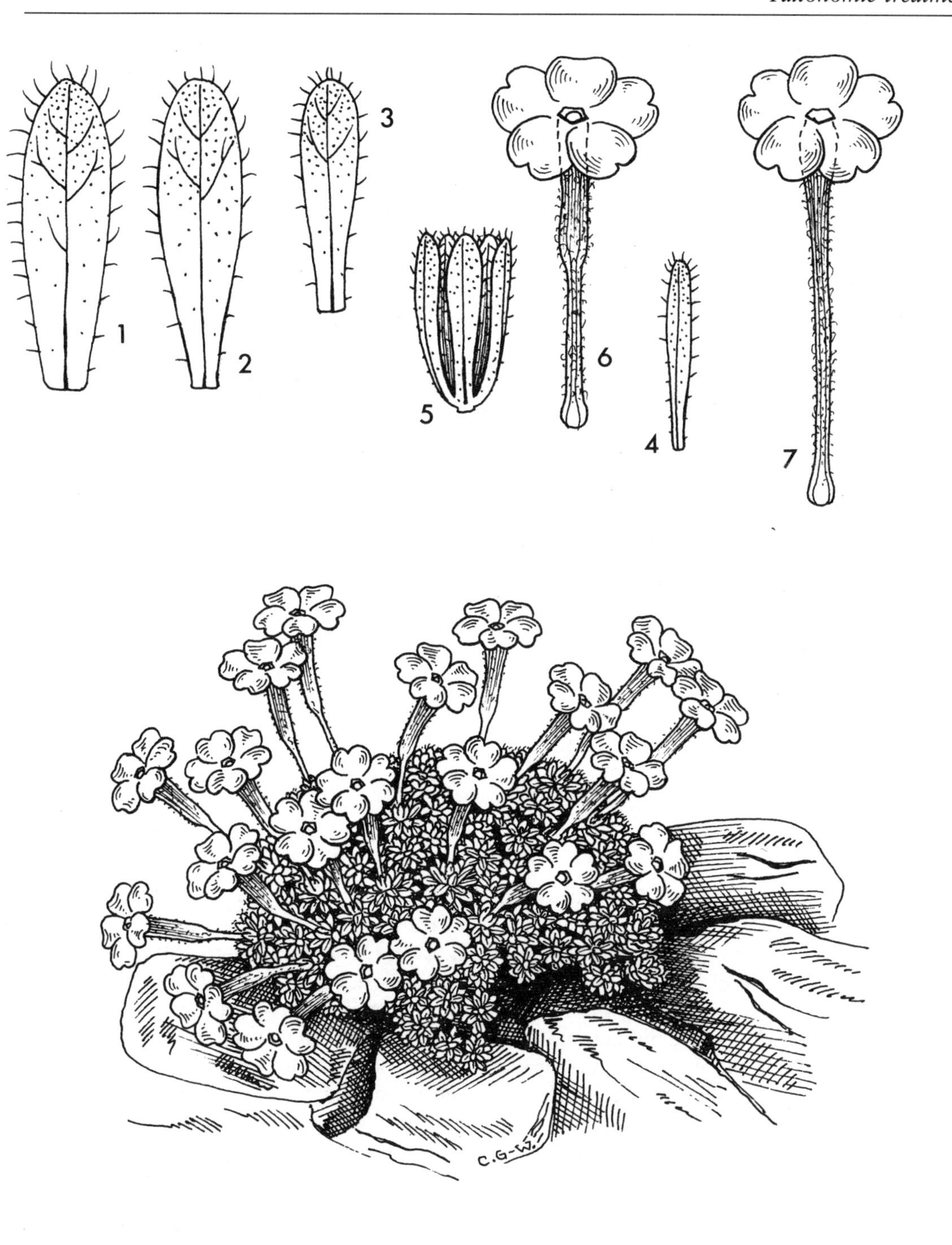

Dionysia haussknechtii, × 2. 1–3 leaves, × 12; 4 bracts, × 12; 5 calyx, × 12; 6 pin-eyed corolla, × 4½; 7 thrum-eyed corolla, × 4½.

marginal fringe; veins, except the midrib, obscure. *Flowers* solitary, sessile. *Bracts* 1, linear-oblong, 2.5 mm long, covered in glandular-hairs. *Corolla* yellow, glandular-hairy outside; tube c. 15 mm long; limb 6–7 mm diameter, divided into broad-obcordate lobes, shallowly emarginate. *Anthers* 1 mm long, inserted in the throat or near the middle of the corolla-tube. *Style* reaching just below the throat or the middle of the corolla-tube. *Capsule* with c. 3 seed. (Plate 23)

DISTRIBUTION. Iran: Luristan Province, Shuturun Kuh, near Aligudarz & Gahar; altitude unknown.

HABITAT. Probably on limestone cliffs and flowering in April and May.

Dionysia haussknechtii was collected on Shuturun Kuh in Luristan Province, Iran, on several occasions between 1895 and 1908. It is an attractive yellow-flowered species closely related to *D. lamingtonii*, differing in the calyx which is split to the base and in the leaves which have a fringe of articulated hairs, but which are otherwise covered in minute capitate glands. *D. michauxii* also has a similarly split calyx, but in this species the leaves, bracts and calyces are densely covered all over in long articulate hairs.

Since Strauss's day *D. haussknechtii* has scarcely been re-collected. Koelz found it on an untraceable mountain called Gahar in 1941 and Jim Archibald collected and photographed it (pl. 23) in 1966.

Unfortunately this fine species is not in cultivation, though it would certainly be the rival of both *D. lamingtonii* and *D. michauxii*, both of them with the tightest of cushions and the brightest of yellow flowers.

24 *Dionysia lamingtonii*

Dionysia lamingtonii Stapf in Kew Bull. 1913: 43 (1913). Type: Iran, Bakhtiari, July 1912, *Lamington* s.n. (holotype K.).

DESCRIPTION. Plant forming a dense greyish cushion (to 20 cm diameter in the wild), efarinose. *Stems* columnar, short-branched, becoming woody below, above covered by closely overlapping dead leaves. *Leaves* oblong to oblong-spathulate or spathulate, 2.5–3.5 mm long, 1–1.5 mm wide, the apex obtuse to subacute, or slightly acuminate, the margin entire, covered in long articulate hairs beneath which become thinner or are absent towards the apex, above glabrous, but hairy along the margin; veins prominent, flabellate, in the upper non-hyaline third. *Flowers* solitary, sessile. *Bracts* 1, linear-spathulate, 1–1.25 mm long, pubescent like the leaves. *Calyx* tubular-campanulate, 1.75–2.2 mm long, divided for two thirds of its length into oblong, obtuse lobes, densely hairy in the lower half on the outside and along the lobe-margins. *Corolla* yellow, sparsely pubescent outside; tube 10–14 mm long; limb 5–6 mm diameter, the lobes suborbicular to obovate, distinctly emarginate. *Anthers* c. 1–2 mm long, inserted in the throat or just below the middle of the corolla-tube. *Style* reaching one-third to half-way, or to the throat, of the corolla-tube (Plates 24–6).

DISTRIBUTION. W. Iran: Zagros Mountains, Bakhtiari region, W of Shahreza; altitude 2850–2950 m.

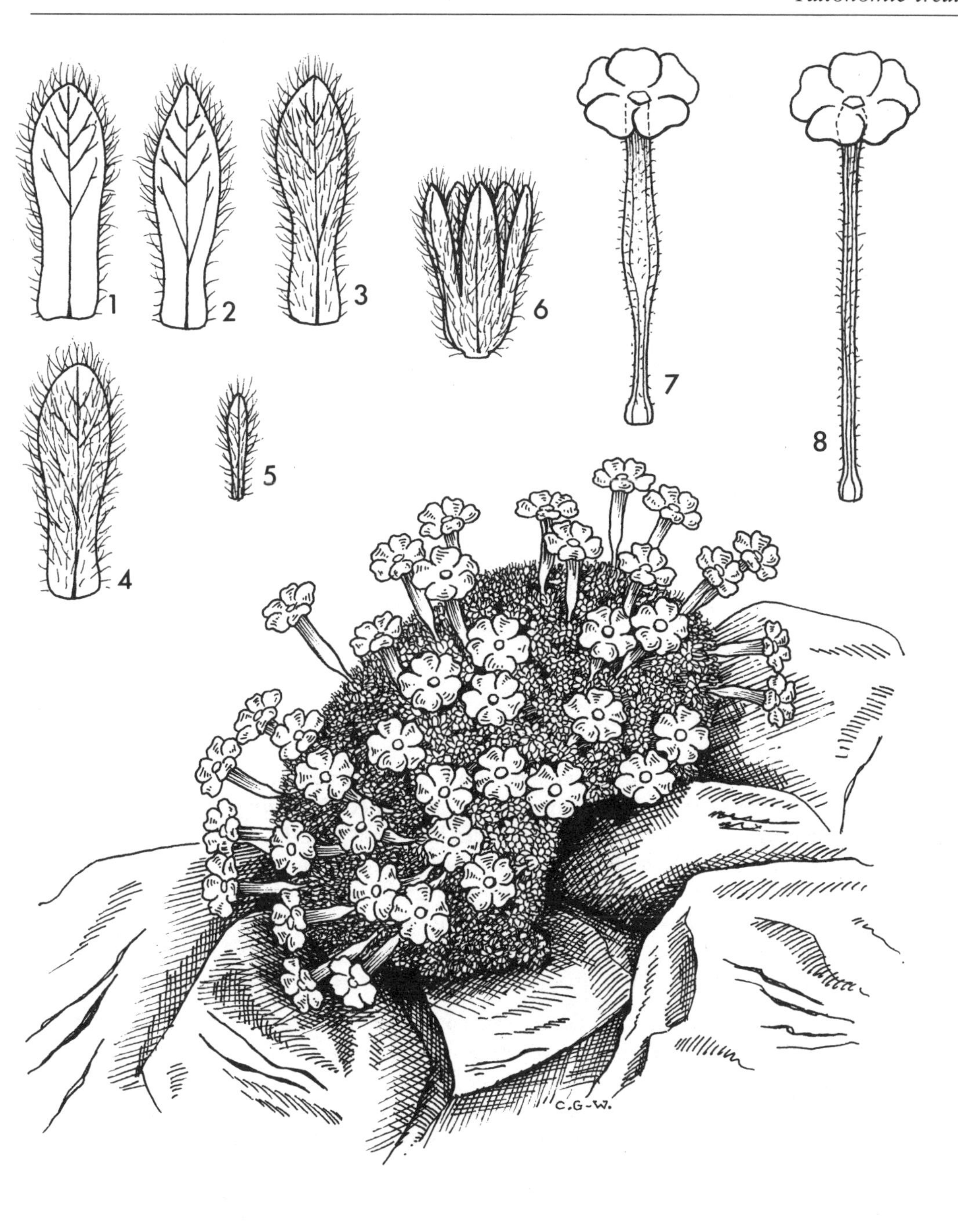

Dionysia lamingtonii, × 2. 1–2 leaves above, × 12; 3–4 leaves below, × 12; 5 bracts, × 12; 6 calyx, × 12; 7 pin-eyed corolla, × 4½; 8 thrum-eyed corolla, × 4½.

HABITAT. Crevices of SSW-facing limestone cliffs in full sun or partial shade. Flowering in the wild April–May.

Dionysia lamingtonii was first collected in the Bakhtiari Mountains (C Zagros Mountains) of Iran by Lord Lamington on the 3rd July 1912 and it was subsequently described by Otto Stapf in 1913. The description was based upon rather scanty material and only a brief floral description was possible – the flower colour was unknown!

Per Wendelbo, in his monograph of the genus *Dionysia* (1961), was not able to elucidate much further as no material had been collected in the intervening years. In his monograph Wendelbo remarks – '... *D. lamingtonii* seems to be a good species, which must be associated with *D. haussknechtii*, *D. sawyeri* and *D. michauxii*, but differing in the thinner texture of the leaves. In *D. michauxii* and *D. haussknechtii* the calyx is divided to the base, while in the other two it is less deeply divided. *D. sawyeri* differs from *D. lamingtonii* in the larger leaves and in the conspicuously ciliate margins to the leaves, calyces and bracts'.

One of the problems was the precise location of Lamington's original find – no other information was given other than the 'Bakhtiari country', which covers a large area. This remained the position until 1973 when Professor T. F. Hewer came across a yellow *Dionysia* in the Zagros Mountains 75 km to the west of the town of Shahreza. Examination of the Hewer specimens at Kew showed that it closely matched the type specimen of Lord Lamington and so represented only the second collection of this apparently rare and little known species. A full description of *D. lamingtonii* together with details of its re-collecting were presented in the *Kew Bulletin* (Grey-Wilson, 1974) together with a detailed figure of the species.

It is likely that this recent collection was gathered in, or very near to, the type locality. Professor Hewer saw and photographed *D. lamingtonii* in full flower in early April and his collection includes both pin- and thrum-eyed corolla forms.

In general details *D. lamingtonii* would seem to come closest to *D. michauxii*. It differs primarily in the calyx which is divided for only two-thirds of its length (not to the base) and in the leaves being practically hairless above. In *D. lamingtonii* the leaves and calyx-lobes are almost devoid of hairs towards the top and look quite different from those of *D. michauxii* which are densely hairy.

In my early book on Dionysias (1970) I remarked of *D. lamingtonii* that – 'It is unlikely that this species will be collected again in the near future, unless it is come upon by accident. Most of the Bakhtiari country consists of low foothills and mountains separated by long flat-bottomed valleys, and there are thousands of likely localities for it. One is just as likely to come across a new species in such country'.

It therefore came as a considerable surprise to me when Professor Hewer sent to Kew specimens which proved, after a detailed study, to be *D. lamingtonii*. He had indeed come upon this species by chance, having ventured west from the city of Esfahan with friends for a day out and picnic. It is interesting to relate that since 1970 six new species have been discovered in the Zagros Mountains, *D. esfandiarii*, *D. iranshahrii*, *D. lurorum*, *D. sarvestanica*, *D. termeana* and *D. zagrica*. There is every reason to expect more such finds in these mountains in the future.

Detritus collected by Professor Hewer, together with herbarium specimens, yielded a number of seedlings (H. 1909) and all the plants in cultivation have come from this source.

Seed from the Hewer collections has produced beautiful, compact and rather uniform offspring. *D. lamingtonii* like *D. michauxii* forms a very compact rounded cushion. The cushions are grey-green and bear very neat small flowers, like tiny funnels, in great profusion. In the short time it has been in cultivation *D. lamingtonii* has proved a real jewel, but it cannot be classed amongst the easiest to cultivate and maintain and is still relatively rare in collections – it cannot, however, be said to be as temperamental in cultivation as its close cousin *D. michauxii*, which like most of the very hairy silvery-grey cushioned Dionysias, has proved a brute to grow well.

Dionysia lamingtonii is a slow-growing species perhaps reaching 5 cm in diameter in 3–4 years. The largest cushion in cultivation, after 14 years, was 15 cm diameter. The cushions are grey and very hairy and as a result are very prone to fungal attacks during damp dull autumn and winter weather. The plants, which form mounded cushions, often look dead during the winter, but remarkably, generally come into life again as spring approaches.

In flower *D. lamingtonii* is very beautiful. The flowers are small, but elegant and produced in profusion all over the neat cushion. At a glance they may be mistaken for *D. tapetodes*, but the cushions are much greyer, more mounded and the corolla-tube is more slender, hairy, not glabrous, on the outside.

Cuttings afford the only means of propagation at present as no seed has been set in cultivation. In fact both pin- and thrum-eyed plants are in cultivation so that it should be possible to get some seed. Cuttings may give a success rate of 50–60%, but sometimes less. The best time is May and early June, but it is not always easy to get sufficient and suitable material for cuttings. Cuttings from young plants appear to root more readily than those of older, long-established plants. Cuttings have been rooted in a gritty compost, or pumice, and more recently Senagel has given remarkably good results.

Dionysia lamingtonii is not easy to keep in cultivation. Plants can die without warning. Older plants become increasingly prone to fungal and insect attacks, but despite all this the species is well-worth persevering with; it has been the focus of attention at more than one Alpine Garden Society show. In cultivation plants flower between February and April.

25 *Dionysia michauxii*

Dionysia michauxii (Duby) Boiss., Diagn. Plant. Orient. Nov. 1(7): 67 (1846).

Gregoria michauxii Duby, Primulaceae in DC. Prodr. 8: 46 (1844). Type: Iran, *Michaux* s.n. 1783–84 (holotype G; isotypes K, P).

Primula michauxii (Duby) Kuntze, Revisio Gen. Pl. 2: 400 (1891).

Description. *Plants* forming dense rounded silvery-grey cushions, efarinose. *Stems* short-branched, crowded and columnar, becoming woody below, densely covered in overlapping dead leaves and leaf-remains. *Leaves* oblong to oblong-

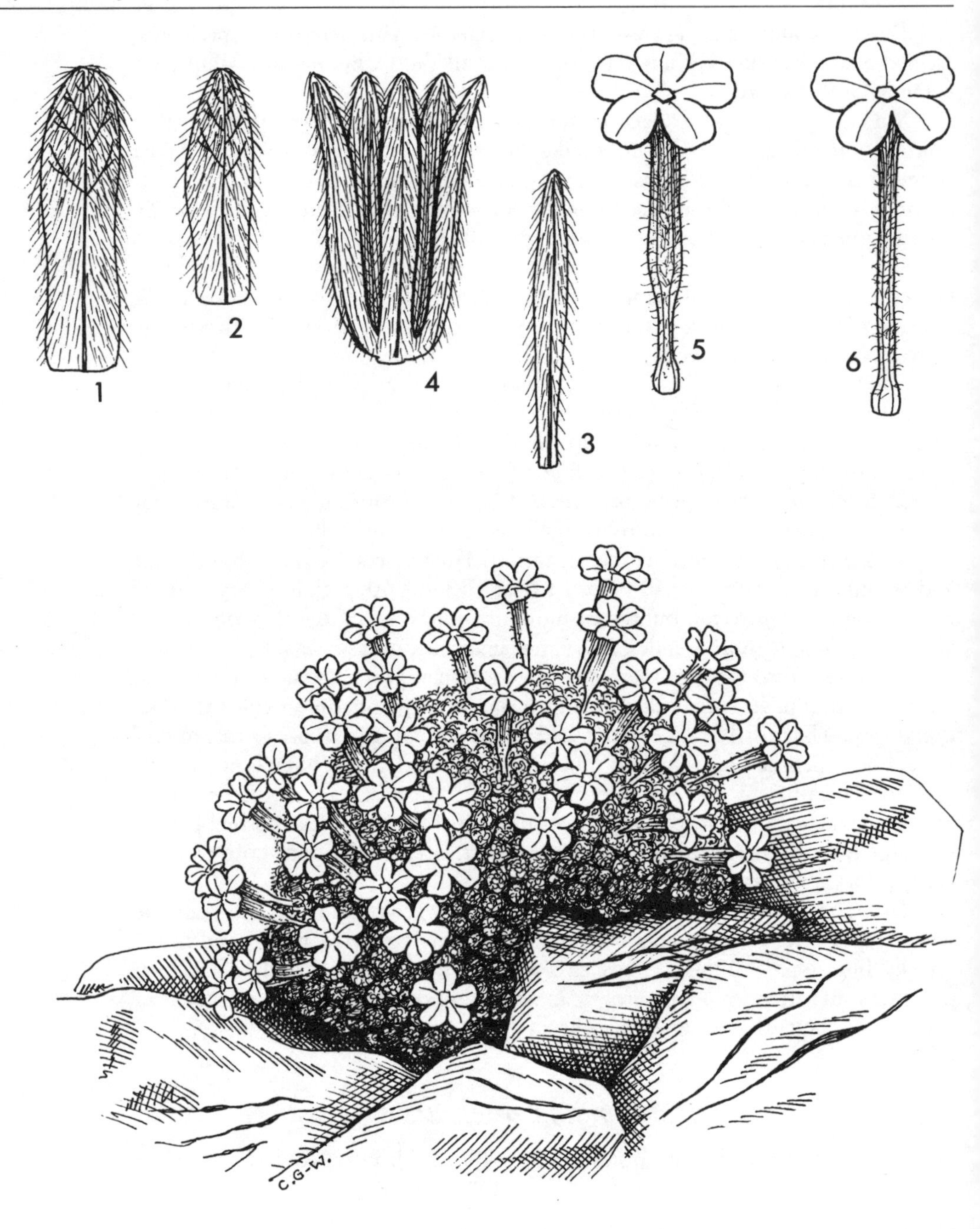

Dionysia michauxii, × 2. 1–2 leaves, × 15; 3 bracts, × 15; 4 calyx, × 15; 5 pin-eyed corolla, × 4½; 6 thrum-eyed corolla, × 4½.

51 *Dionysia viscidula*. Photo. E. G. Watson.

52 *Dionysia viscidula*. Photo. E. G. Watson.

53 The limestone Gorge of Darrah Zang, NW Afghanistan, on whose cliffs are found *D. afghanica, D. lindbergii, D. microphylla, D. tapetodes* and *D. viscidula*, together with a number of other interesting chasmophytic plants. Photo. C. Grey-Wilson.

54 Limestone cliff at Darrah Zang, NW Afghanistan, with numerous cushions of *D. afghanica* – the type locality. Photo. C. Grey-Wilson.

55 *Dionysia afghanica*. Photo. E. G. Watson.

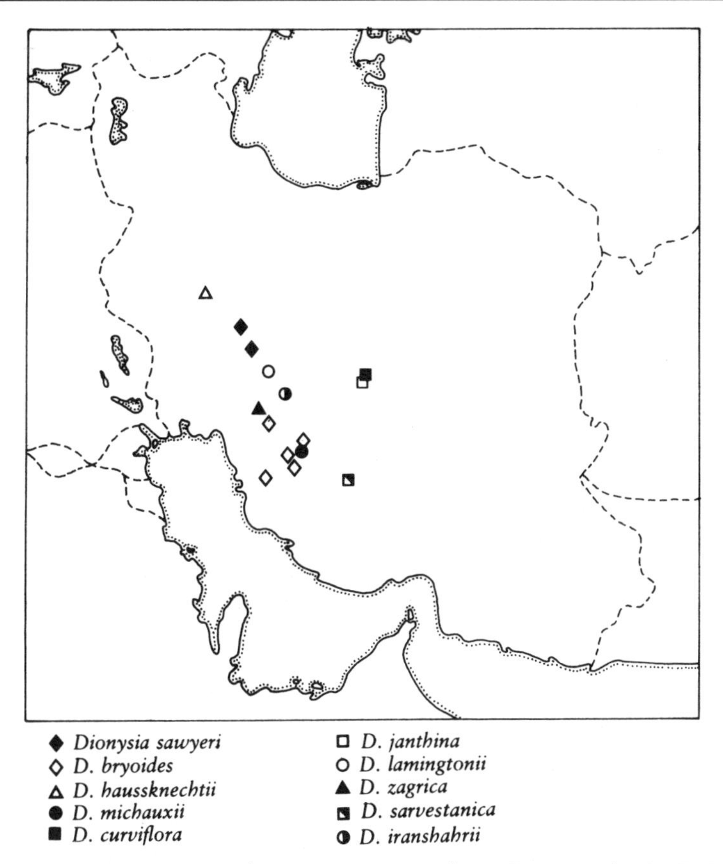

Map 5 Distribution of section *Dionysia* subsect *Bryomorphae* in W. Iran.

elliptic, obtuse to subobtuse, entire, covered on both surface with reflexed to spreading articulate hairs; veins rather obscure. *Flowers* solitary, sessile. *Bracts*, linear-oblanceolate, 3 mm long, acute, hairy like the leaves. *Calyx* tubular-campanulate, divided to the base into linear-elliptic, acute or subacute lobes, covered in reflexed articulate hairs. *Corolla* yellow, hairy outside; tube c. 10 mm long; limb 6–7 mm diameter, divided into obovate, very slightly emarginate, lobes. *Anthers* c. 1.5 mm long, inserted in the throat or near the middle of the corolla-tube. *Style* reaching the throat or about half-way along the corolla-tube. *Capsule* with c. 3 seeds (Plates 27–8).

Distribution. SW Iran: Fars Province, Kuh-i-Bamu, E of Shiraz; altitude 2200–2250 m.

Habitat. Growing on E and SE facing limestone cliffs in sun or partial shade; flowering during April and May.

Dionysia michauxii is without doubt one of the most beautiful and compact of all species of *Dionysia*. The beauty of this species was first revealed to alpine gardeners by the pictures of glowing yellow balls of flowers nestling on cliffs in southern Iran which were taken by Jim Archibald. He had visited the type locality, Kuh-i-Bamu, in 1966 and found the plants in full flower.

Dionysia michauxii was first discovered in 1783–84 by André Michaux in whose honour the plant was named by Duby – he had placed it in the genus *Gregoria*, but it was not until 1846 that Boissier transferred it to *Dionysia*.

This species is found on one mountain only, Kuh-i-Bamu, to the east of the city of Shiraz. Like many other species in the genus, *D. michauxii* appears to be very restricted in the wild. Jim Archibald comments – '... *D. michauxii* does not grow on the apparently 'suitable' shaded basalt cliffs, but only on the limestone outcrop of Kuh-i-Bamu ... south-east facing cliffs, but usually shaded after the early morning by outcropping rocks or overhangs'.

The species has had a chequered history in cultivation. Seed was first introduced into cultivation by Giuseppi in 1932 and a plant was shown in 1935. However, all the plants were quickly lost. In 1939 Dr Peter Davis introduced further seed and a plant from this source was shown in flower in 1944 and given a Cultural Commendation by the Alpine Garden Society. However, all the plants from this source also perished. A further introduction was made in 1966 by Jim Archibald and this has faired better; plants are still in cultivation today from this source. It is easy to see why the earlier introductions failed to establish this exquisite species in cultivation, for it is proving one of the most difficult of all Dionysias to please. Plants are temperamental about repotting, dislike more than the minimum of watering and die away 'in minutes' if accidentally watered overhead. Jack Elliott wrote in 1968 – 'What a brute! I have germinated three, but have lost them all after repotting, in spite of trying different methods'. Peter Edwards wrote at the same time – 'a neat silver cushion plant which needs hardly any watering in cultivation, even in spring and summer. Any over-watering creates a loose cushion which is not its natural form. One of the more difficult to keep growing, although not difficult to propagate ...'.

Today only a single clone (pin-eyed) is in cultivation (JCA); though kept alive by the regular progagation of cuttings it is not easy to keep going. Plants reach about 3 cm diameter in 2 years and 5–7 cm in 3 years. The oldest plant in cultivation (some 10 cm diameter) is 11 years old. *D. michauxii* forms very tight silvery-grey, very hairy cushions that are particularly prone to botrytis attack, especially during the autumn and winter – for this reason plants should never be watered directly, but instead the plunge is kept moist.

Cuttings taken in May and June, and carefully guarded against botrytis, can give a good result – 70–90% success has been recorded, but this is perhaps exceptional and few can claim regular success. As Henrik Zetterlund remarks – 'Can't say I grow this species – I kill it. We are dependent on Eric Watson who regularly supplies us with new plants'.

Plants occasionally suffer aphid attacks and these should be watched for, especially in the early spring.

In cultivation plants flower during February and March.

26 *Dionysia curviflora*

Dionysia curviflora Bunge in Bull. Ac. Imp. Sci. St.-Petersb. 16: 562 (1871). Type: Iran, Yazd, Schir Kuh, *Buhse* 1352 (holotype G; isotypes B, P).

Primula curviflora (Bunge) Kuntze, Rev. Gen. Pl. 2: 400 (1891).

Dionysia bryoides sensu Boiss. et Buhse, Aufz. Reise, Transkaukas. Pers. Pflanz. (Moscow): 145 (1860), non Boiss.

Dionysia ianthina sensu Giuseppi in Bull. Alp. Gard. Soc. 2: 2 (1933), non Bornm. et Winkl.

DESCRIPTION. *Plants* forming dense, rather flat, grey or green cushions to 60 cm diameter, though often less, efarinose. *Stems* short-branched, columnar, becoming woody below eventually, covered in dead overlapping leaves and leaf-remains. *Leaves* oblong to oblong-oblanceolate or oblong-spathulate, 2–3 mm long, 0.8–1 mm wide, subacute, entire, with long articulate hairs along the margin and in the upper half above, glabrous beneath; veins obscure. *Flowers* solitary, sessile, heterostylous. *Bracts* 2, linear-lanceolate, 2.5–3 mm long, acute, hairy like the leaves. *Calyx* narrow-campanulate, 3 mm long, divided for about two-thirds of its length into linear-elliptic acute lobes, hairy along the margins and inside, but glabrous outside. *Corolla* pink, rarely whitish, often with a yellowish 'eye', glabrous; tube 11–12 mm long; limb c. 6 mm diameter, divided into narrow- to broad-obcordate lobes, deeply emarginate. *Anthers* c. 1 mm long, inserted in the throat or middle of the corolla-tube. *Style* reaching the middle or the throat of the corolla-tube. *Capsule* containing 4–5 seeds (Plates 29–30).

DISTRIBUTION. Iran: Yazd Province, Schir Kuh – not known elsewhere; altitude 2700–4000 m.

HABITAT. Growing in crevices in N- and W-facing basalt or volcanic cliffs, in shade and partial shade; flowering during March, April and May.

Dionysia curviflora was first collected by Buhse on Schir Kuh near Yazd (Yasd) in 1849. This beautiful species was at first mistaken for *D. bryoides* by Boissier and Buhse (1860) but its true identity was revealed by Bunge in his monograph of 1871.

Dionysia curviflora is one of two species in the area around Taft in central Iran, the other species being *D. janthina*. The two species are very similar in some respects and, as a result, they have given rise to a good deal of confusion over the years. It may help to say from the outset that Giuseppi's *D. ianthina* (spelt with an *i*) referred to in early volumes of the Alpine Garden Society Bulletin is a synonym of *D. curviflora* and therefore distinct from *D. janthina* Bornm. & Winkl. Much of this confusion was started by Bornmüller who discovered *D. janthina* but, when describing it, distinguished it from *D. curviflora* by stating that the latter species had yellow flowers*, though he had never seen it. Later Giuseppi, probably following Bornmüller, states that *D. ianthina* (ie *D. curviflora*) has

*In his original description of *D. curviflora* Bunge makes no mention of the flower colour.

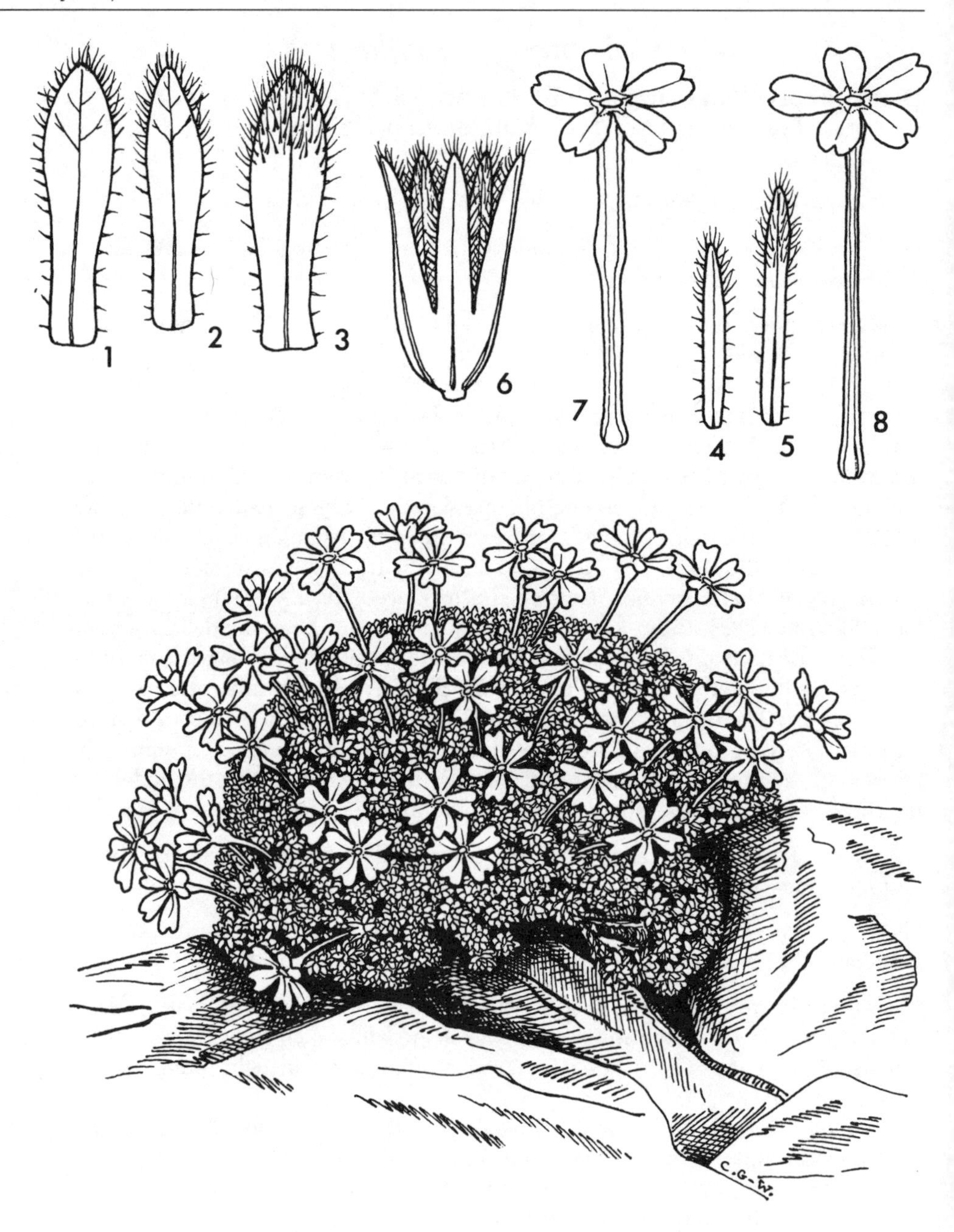

Dionysia curviflora, × 2. 1–2 leaves below, × 12; 3 leaf above, × 12; 4–5 bracts, × 12; 6 calyx, × 12; 7 pin-eyed corolla, × 4½; 8 thrum-eyed corolla, × 4½.

yellow flowers. Clearly he was confused here because he, in the company of E. K. Balls in 1932, had seen this species flowering in the wild. Balls wrote then – 'Here on Schir Kuh *Dionysia* justifies all the praise that has been, or can be, lavished on that wonderful family ... *D. ianthina* (curviflora) sprang out at us from the rocks by the trackside. Then up and up from the sheer cliffs above, blobs and patches of pink in many shades glowed down upon us ...'. Both *D. curviflora* and the true *D. janthina* are now known to have pink flowers, indeed no yellow-flowered species has ever been found in the vicinity.

The next confusion between the two species lies in the colour of the cushions. Whereas *D. janthina* always has silvery grey cushions, *D. curviflora* may have, as I see it, grey, or more usually green, cushions. It has been assumed in the past that the two species could be distinguished on cushion colour, but this is not always possible without looking at the botanical characters. *D. janthina* is not in fact known from Schir Kuh itself, but from neighbouring mountains just S of the town of Taft. Both Paul Furse and Brian Mathew record the grey and green cushions of *D. curviflora* on Schir Kuh. Paul Furse wrote – 'On the Schir Kuh cliffs the green and grey cushions are mixed up, and most of them are actually grey, but the specimens all prove to be *D. curviflora*. I think that the age of the plant, the heat, and perhaps other things affect the colour; the plants in the most sun may be a little whiter, and the larger plants whiter, though I have no certainty about this'.

Jim Archibald wrote in 1968 – 'One geographical point – although everybody refers to the great range near Yazd (Yezd) as Schir Kuh, this is only one mountain. Quite a large chunk of country is covered by mountains, starting from Kuh-i-Fakhrabad in the SE and extending NW. As no collecting, as far as I know, has been done either SE or NW on the actual massif centering on Schir Kuh, we have not the faintest idea of the distribution of *D. curviflora* and *D. janthina*'.

The differences between the two species are fairly clear-cut. In the herbarium specimens the appearance of the two is quite striking; whereas the cushions of *D. janthina* are seemingly covered in hairs, those of *D. curviflora* appear almost glabrous on the outside, although the insides of the leaf rosettes are very hairy. This is because both the outer and inner surfaces of the leaves are pubescent in *D. janthina*, whereas only the inner surfaces of the leaves is hairy in *D. curviflora*. Per Wendelbo wrote (a letter in 1969) – 'As regards the problem of separation of *D. curviflora* and *D. janthina* I can just refer to my monograph. I remember that I thought for sometime that the two were synonymous, but after a closer study I found that there were differences in leaf-shape correlated with pubescence ...'.

The leaves of *D. curviflora* have a narrower, more acute apex; they are glabrous beneath but hairy on the margin and on the upper third above. The calyx is glabrous outside, but hairy inside and on the margins of the lobes. In contrast, the leaves of *D. janthina* have a broader, more obtuse apex and they are hairy beneath and on the margin, and on the upper half above. The calyx is hairy outside and on the margin of the lobes, but glabrous inside. These characters are the opposite in each instance.

There is little difference to be noted in the corollas of the two species, although the lobes of *D. janthina* do seem to be generally broader and more overlapping. However, the corolla-lobes of *D. curviflora* do vary considerably from plant to plant. Both species have corollas with a pale to deep yellow 'eye'. E. K. Balls (1934) wrote that – 'The flower ... is very like the single flower of

Primula farinosa. Usually pink with a lighter centre and a yellow 'eye', we found forms varying from pure white to bright pink ... These blooms, growing sessile over the bright green cushions of tiny, packed rosettes laid flat against the vertical faces of the rocks, are really wonderful. From just over 9,000 feet to the very summit over 13,000 feet, the *Dionysia* grows in all the crevices in the cliffs, whether on volcanic rock or granite, generally preferring shade, but also in full sun'.

As already indicated *D. curviflora* does not grow on the same mountains as *D. janthina*, though both grow in the same general vicinity. They may also inhabit different rock types. Balls (1934) records that *D. curviflora* grows on granite or volcanic rock. *D. janthina* in contrast, together with most other *Dionysia* species, appears to inhabit limestone mountains.

Dionysia curviflora has had by far the longest history in cultivation of any *Dionysia* species and is generally considered amongst the easiest to cultivate. This fine species has been in continuous cultivation since 1932 from seed introduced by Giuseppi, which he had collected on Schir Kuh in the company of E. K.Balls. Plants from this source are still in cultivation to this day, though they cannot be said to be as floriferous as more recent introductions. Seed was also collected in 1939 by Peter Davis and certainly plants from this source are more free-flowering. More recent collections, particularly those of Paul Furse, 1962; Brian Mathew, 1963 (Bowles Scholarship Memorial Expedition); and Jim Archibald, 1966, have introduced a variety of forms, most of which are far more free-flowering than those of Giuseppi. Unfortunately, although cultivated plants vary in colour from pale to quite deep pink, the white form mentioned by E. K. Balls has never been brought into cultivation; a pity as apart from a rare form of *D. involucrata* there is no white-flowered *Dionysia* in cultivation.

Peter Edwards wrote in 1969 – 'My largest plant is from a pre-war clone and does not flower, but is amiable and continues to grow. Not difficult and can be easily propagated by tearing rooted pieces from the cushion, or single rosettes, in the spring ...'.

It was the usual practice with *D. curviflora* to pot on the plant and pot entire into progressively larger pots, but apart from preventing possible damage to the cushions there is no reason why they should not not be potted on in the more conventional way. If the first method is resorted to it is probably advisable to knock the bottom off the pot carefully first before placing it in another, larger pot.

Plants grow fairly rapidly from seeds or cuttings, reaching 10 cm diameter in 4 years, but slowing down thereafter – 8 year old cushions are generally 15–20 cm diameter. This depends to some extent on the particular clone for some are only 20 cm diameter after as many years. The largest cushion in cultivation measures about 24 cm across.

Plants form rather low mounds, symmetrical in the early years but often becoming somewhat misshapen in time. Clones vary greatly in flowering habit. The early Giuseppi plants tend to be shy flowers, although in some years they make a bolder show – the reasons for this are not clearly understood but may reflect the conditions the previous summer or winter. Clones raised from Archibald seed (originally introduced under JCA 2800) are very much more floriferous and although the Giuseppi clone is of historical importance it has been surpassed by the later introductions.

Dionysia curviflora will succeed equally as well in a gritty compost as in a block of tufa in the alpine house. Tufa-grown plants are rather slow to make a good tight cushion. Those growers who have been brave enough to try a plant outdoors on a raised scree-bed report that little success has been achieved, the plants soon succumbing to winter dampness, even when protected by a sheet of glass.

Plants may suffer botrytis attacks, especially during the autumn and winter and this may damage part of a cushion causing die-back, but this rarely infects the entire plant and they recover once the spring arrives.

Cuttings are relatively easy to root and a success rate of 50–90% can be expected. However, growers are not generally agreed on the ideal time to take cuttings. Jack Wilkinson states that – 'single rosette cuttings in sharp sand, taken in May, root very well, taking about 5–6 weeks'. Brian Burrow strikes cuttings between 'May–September with an 80–90% success rate'. Henrik Zetterlund, on the other hand says that – 'Late July is definitely the best time to root this species'. This apparent discrepancy almost certainly reflects cutting techniques and local conditions.

Both pin- and thrum-eyed plants are well established in cultivation and it is rather strange that more growers don't collect and germinate their own seed. Seed is often difficult to observe and can be easily overlooked. However, Eric Watson has raised numerous plants from his own seed (originating in the first instance from JCA 2800). Certainly it is beneficial to produce home grown seed from time to time and establish a wider variety of clones. Some of the most floriferous, with deep pink, broad-petalled, flowers should be given cultivar names – it is inconsistent that this most widely grown of any *Dionysia* species should have no named cultivars, whereas the equally popular *D. aretioides* has had several selected over the past twenty years. This may be a reflection of its unreliable flowering habits. However, some of the more floriferous clones surely deserve some recognition.

Dionysia curviflora then, is not just a connoisseurs plant. It is certainly a more-or-less trouble free and reliable species, ideal for the first-time grower to try.

27 *Dionysia janthina*

Dionysia janthina Bornm. et Winkl. in Bull. Herb. Boiss. 7: 70, t. 2 fig 2 (1899). Type: Iran, Yazd, Schir Kuh, *Bornmüller* 3869 (holotype B; isotypes BM, E, JE, K).

Primula janthina (Bornm.) Bornm. loc. cit. 73 (1899).

Dionysia ianthina Knuth, Dionysia in Pax & Knuth, Pflanzenriech 4 (237): 164 (1905).

DESCRIPTION. *Plants* forming silvery-grey cushions very like *D. curviflora* in overall appearance and general characteristics, but varying in the following features. *Leaves* oblong-spathulate, 2–3 mm long, 1–2 mm wide, obtuse to somewhat retuse, with long articulate, spreading hairs along the margin and dense, adpressed hairs beneath and on the upper half above. *Bracts* linear-

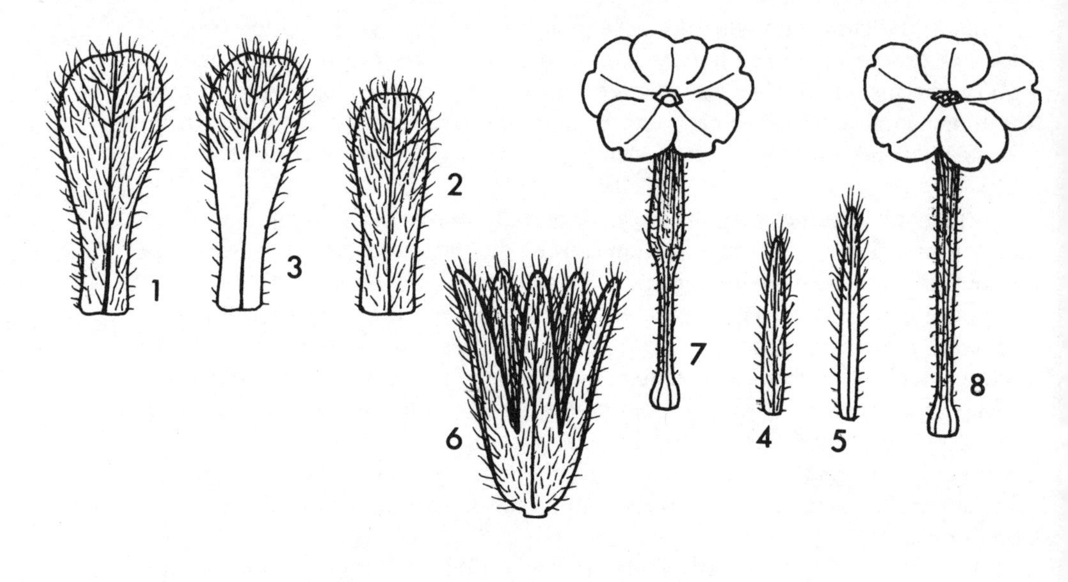

Dionysia janthina, × 2. 1–2 leaves below, × 12; 3 leaf above, × 12; 4–5 bracts, × 6; 6 calyx, × 6; 7 pin-eyed corolla, × 4½; 8 thrum-eyed corolla, × 4½.

lanceolate, hairy as the leaves. *Calyx* campanulate, 2.5–3 mm long, glabrous inside, but covered in dense adpressed hairs outside. *Corolla* pink, often paler in the centre and with a yellowish 'eye'; tube c. 10 mm long; limb 6–7 mm diameter, divided into broad-obcordate, deeply emarginate lobes. *Capsule* with c. 3 seeds (Plates 31–2).

DISTRIBUTION. C Iran: Yazd Province, near Taft, Kuh-i-Buruk; altitude 1800–2000 m.

HABITAT. Growing in crevices of N- and NW-facing limestone cliffs; flowering in late February, March and early April.

For general remarks on *D. janthina* see notes under its close ally, *D. curviflora* on p. 125.

Dionysia janthina was discovered on Schir Kuh near Taft in C Iran by Bornmüller in 1892. The locality has caused some confusion over the years as Schir Kuh is the type locality of its closest ally *D. curviflora*. In fact it is now recognised that *D. janthina* is not known from Schir Kuh, but from neighbouring mountains. It is probably therefore that Bornmüller used Schir Kuh to cover all the mountains in the vicinity of Taft. Shortly after seeing *D. janthina* in the wild in 1966 Jim Archibald wrote – '*D. janthina* does not grow on Schir Kuh, or on the Kuh-i-Barfkhaneh connected to it. We only know of it immediately S of Taft – it may well grow in greater quantities miles to the NW of Taft ... vertical fissures on NW-facing cliffs at 1800 m on a hill leading up to the Kuh-i-Buruk, south of Taft ...'.

Apart from growing on different mountains in the vicinity of Taft and apparently inhabiting different rock types (see p. 127) *D. janthina* and *D. curviflora* grow at different altitudinal levels; *D. curviflora* grows between 2700 and 4000 m.

Dionysia janthina is, in many ways, an even finer plant than *D. curviflora*, with its silvery-grey cushions and more substantial blooms. It is by no means such an easy and amenable plant in cultivation and has proved far more difficult to maintain and propagate. Indeed it is still a rare plant in collections.

The first plants in cultivation were from seed collected by Jim Archibald in 1966 who wrote shortly after – 'My main excitement is that I have managed to keep two out of three seedlings going of *D. janthina* – giving it *D. michauxii* treatment, but even drier. The winter may be the difficult time as this was collected at a lower altitude than *D. curviflora* ... It is extremely distinct from *D. curviflora* (and all the others) in the seedling stage and much slower to "compact itself" – all are lush and look like Farinosae primulas when they first germinate. The juvenile leaves are very distinct from any *D. curviflora* seedlings (the latter are very variable, of course) – the margins are distinctly crenate-dentate'.

Subsequent history has shown that *D. janthina* is not particularly difficult in cultivation, although not as easy as *D. curviflora*. Plants are reasonably fast-growing, reaching up to 10–12 cm in 4 years, 18 cm in 8 years and 22 cm in 13 years. The plants form low rather soft mounds, paler than those of *D. curviflora* and seldom the silvery-grey cushions of the species as it is seen in the wild.

Young plants have large more open rosettes but as plants mature the rosettes become tighter and the cushions denser. Single rosette cuttings of young vigorous plants root far more readily than those of mature plants. Using cuttings from young plants a success rate of 100% has been recorded, but 60–80% is more usual. Cuttings can be struck at almost any time from June to September, although earlier rooted cuttings are more established by the time winter arrives and are more likely to survive.

Dionysia janthina is a very beautiful plant in flower with clear pink flowers with broad petals. Unfortunately plants, especially the older ones, are not always very floriferous – this also applies to some forms of *D. curviflora*. In cultivation flowers most frequently appear in late autumn/early winter.

Although this species is usually long-lived it is wise to raise batches of cuttings at regular intervals as young plants tend to be more floriferous.

Seed has not, to my knowledge, been produced, but as far as I am aware only the pin-eyed plant is in cultivation.

28 *Dionysia iranshahrii*

Dionysia iranshahrii Wendelbo in the Iranian Journ. Bot. 1(1): 72 (1976). Type: W Iran, Bakhtiari, Kuh Pashmaku, June 1974, *M. Iranshahr* s.n. (holotype Herbarium Ministry Agriculture, Evin, Tehran; isotypes GB, TARI).

DESCRIPTION. *Plants* forming dense silvery-grey cushions, efarinose. *Stems* short-branched, columnar, covered in closely overlapping dead leaves and leaf-remains. *Leaves* narrow-oblong to lanceolate-elliptic, 2–2.5 mm long, 0.5 mm wide, subacute, entire, covered below and on the margins with long spreading articulate hairs, ± glabrous above or with hairs at the apex; veins obscure. *Flowers* solitary, sessile. *Bract* 1, linear-subulate, 2 mm long, hairy like the leaves. *Calyx* narrow tubular-companulate, c. 2.5 mm long, divided for three-quarters of its length into narrow-lanceolate lobes, covered in long articulate hairs. *Corolla* violet, glabrous, heterostylous (only short-styled flower form known); tube c. 11 mm long; limb 5–6 mm diameter, divided into obovate, very slightly emarginate lobes. *Anthers* 0.75 mm long, inserted in the throat of the corolla. *Style* reaching about one third the way up the corolla-tube. *Capsule* with 2–3 seeds.

DISTRIBUTION. W Iran: Bakhtiari Province, Semirom, Kuh Pashmaku; altitude 2650–3000 m.

HABITAT. Probably crevices in limestone cliffs; flowering in April–May.

Dionysia iranshahrii was discovered on the 6th June 1974 by M. Iranshahr, an inveterate collector who has travelled and collected in many parts of Iran, and in whose honour this species was named by Per Wendelbo (1976).

Dionysia iranshahrii is an interesting, but very little known, species of subsection *Bryomophae*, forming dense, hairy, grey cushions. The violet corollas are, in contrast, quite glabrous. The leaves are hairy on both surfaces, thus clearly distinguishing them from the glandular leaves of *D. bryoides* and the larger ciliate-margined leaves of *D. sawyeri* which are glabrous beneath. The other violet- or pink-flowered members of the *Bryomophae*, *D. curviflora* and *D. janthina*, have leaves that are at least glabrous in part but are, in any case, readily

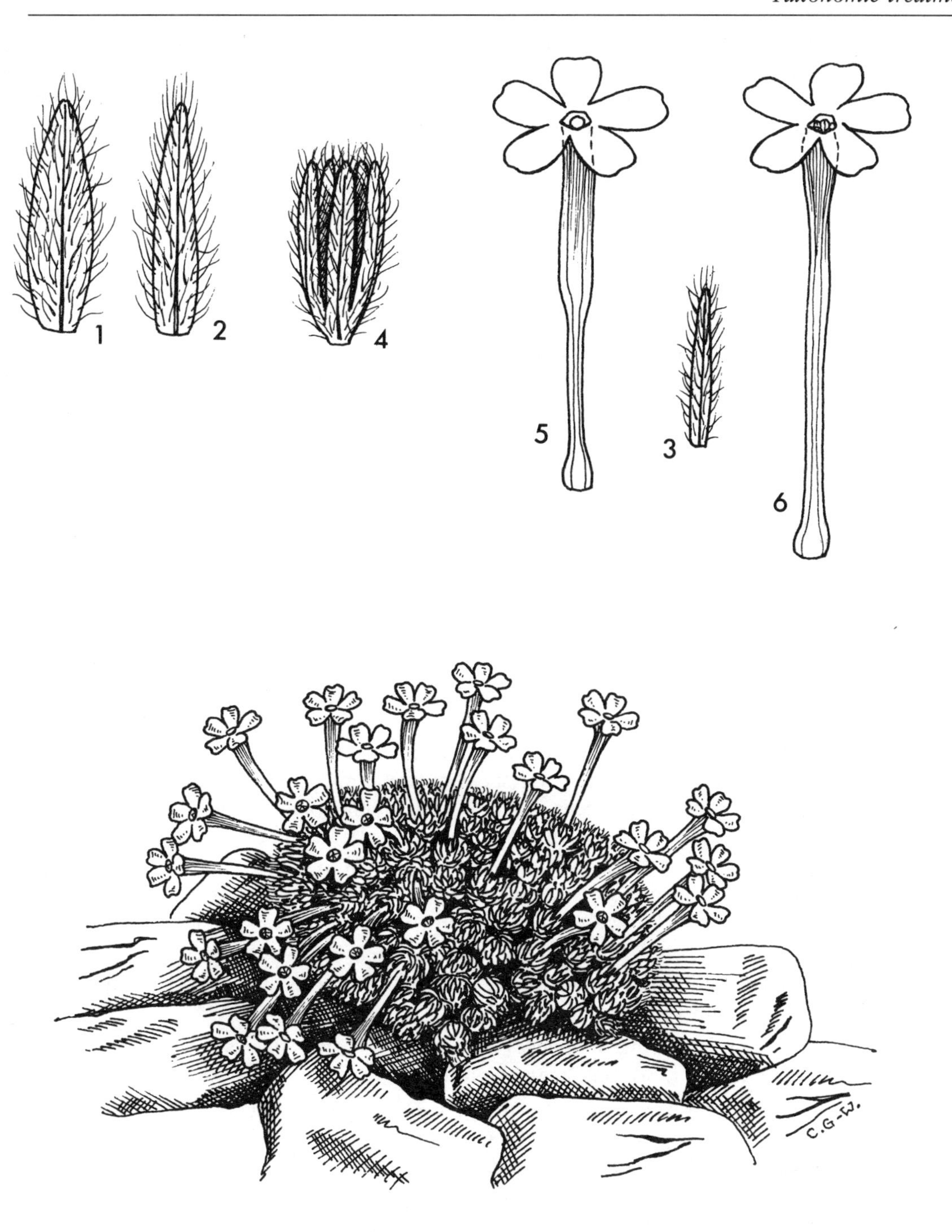

Dionysia iranshahrii, × 2. 1–2 leaves, × 18; 3 bracts, × 12; 4 calyx, × 12; 5 pin-eyed corolla, × 6; 6 thrum-eyed corolla, × 6.

distinguished by their *deeply* emarginate corolla-lobes – those of *D. iranshahrii* are scarcely emarginate.

Dionysia iranshahrii hails from the Zagros Mountains, the home of so many fine species of *Dionysia*, occurring in the Bakhtiari Province – which is also the home of *D. lamingtonii*, *D. archibaldii* and *D. caespitosa*.

Very little is known about the range and natural variation of this interesting species. It has only ever been collected once and has never been introduced into cultivation.

29 *Dionysia bryoides*

Dionysia bryodes Boiss., Diagn. Pl. Or. Nov. 1(7): 66 (1846). Type: Iran, Fars, Kuh Ayub, *Kotschy* (holotype G; isotypes BM, K, W).

Primula kotschyi (Bunge) Kuntze, *loc. cit.*

Primula bryoides Kuntze, Rev. Gen. Plant. 2: 400 (1891).

Diorysia kotschyi Bunge in Bull. Ac. Imp. Sci. St-Petersb. 16: 560 (1871). Type: Iran, Fars, Kuh Ayub, *Kotschy* 406b (holotype LE; isotypes B, P).

DESCRIPTION. *Plants* forming dense, rather flat, deep green cushions (to 30 cm diameter in the wild), efarinose. *Stems* short-branched, ± columnar, becoming woody below eventually, covered in closely overlapping dead leaves and leaf-remains. *Leaves* oval-oblanceolate to oval-spathulate, 2–3 mm long, 0.75–1 mm wide, obtuse to subobtuse, entire, densely covered in minute capitate glands, but without hairs; veins rather obscure. *Flowers* solitary, sessile, heterostylous. *Bracts* 1 or 2, linear-oblanceolate, glandular like the leaves. *Calyx* tubular-campanulate, 2–3 mm long, divided to the base into linear-oblanceolate to linear-spathulate lobes, covered in minute capitate glands. *Corolla* pale to deep pink or violet, generally paler in the centre, minutely glandular outside; tube 9–10 mm long; limb 5–6 mm diameter, divided into obcordate, shallowly emarginate lobes. *Anthers* 1.3–1.5 mm long, inserted in the throat or in the middle of the corolla-tube. *Style* reaching the middle of the corolla-tube or slightly exserted. *Capsule* with 2–4 seeds (Plates 33–34).

DISTRIBUTION. SW Iran: Fars Province, on various mountains including Kuh Ayub, Kuh-i-Dinar, Kuh-i-Made, Kuh Bungi near Abdui, Kuh Khormug near Bushire, Kuh-i-Sabzpuchon near Shiraz; altitude 1800–2800 m.

HABITAT. Growing in crevices on limestone cliffs (often N- or NW-facing), in sun or shade; flowering in April and May.

Dionysia bryoides is a particularly pretty species in the best forms with mid to deep pink flowers. The flowers generally have a white or yellowish 'eye' and the petals vary considerably in width.

This species has had a reasonably long but interrupted history in cultivation. The first plants in cultivation were raised from seed collected by E. K. Balls and Giuseppi in 1932.

Thereafter the species seems to have been lost to cultivation and it was not until it was re-collected by Jim Archibald in 1966 that it once more graced the benches of the alpine shows in Britain. Further seed was collected by Tom Hewer in the summer of 1976. All the plants in cultivation today seem to have come originally from these two sources – Archibald and Hewer.

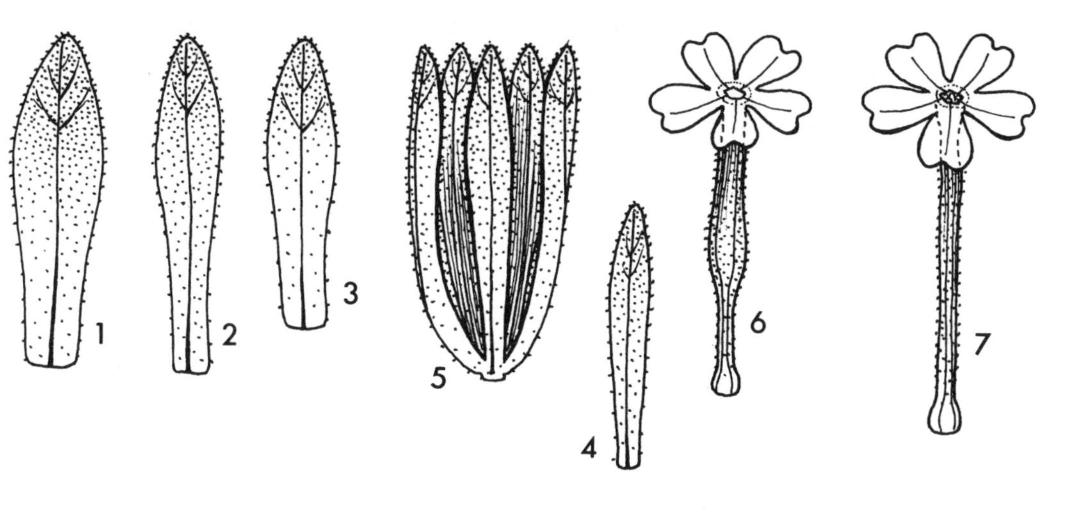

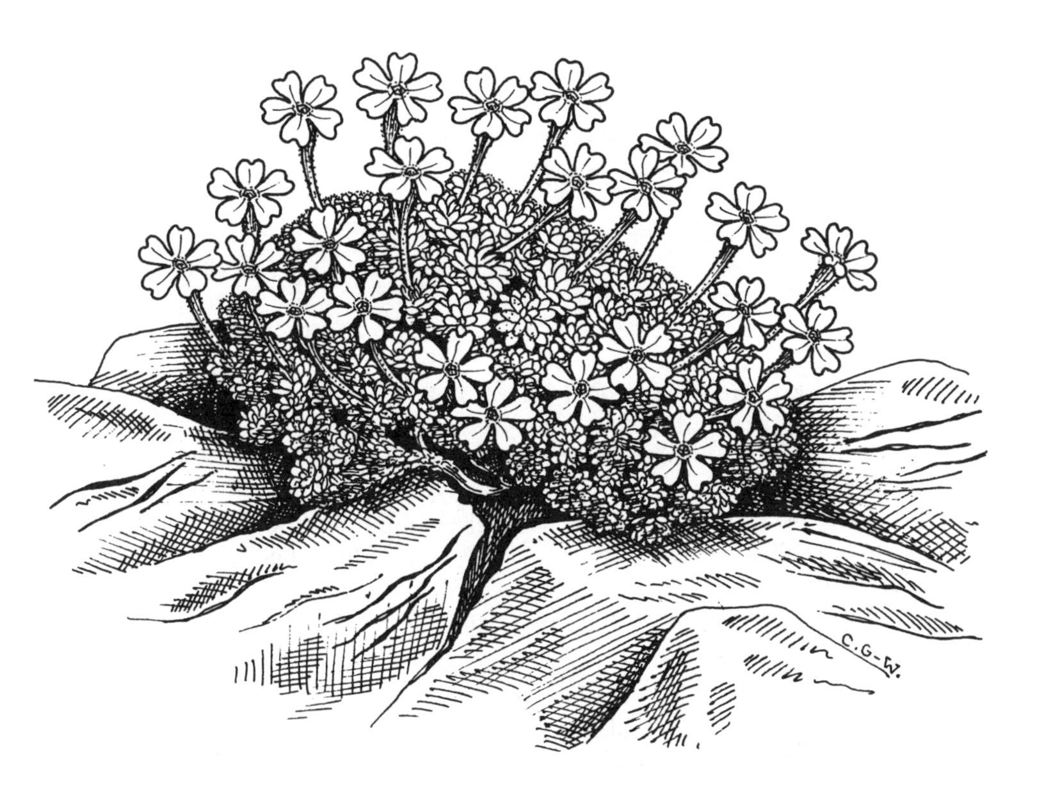

Dionysia bryoides, × 2. 1–3 leaves, × 18; 4 bracts, × 18; 5 calyx, × 18; 6 pin-eyed corolla corolla, × 4½; 7 thrum-eyed corolla, × 4½.

It has to be said that *D. bryoides* has proved far from easy in cultivation and few specimens have reached any age, or indeed any size. The reasons for this are not absolutely clear. *D. bryoides* is a rather unhairy species, the whole plant instead being covered in tiny glands, giving it a sticky feel. On the whole it is the hairy-cushioned species that are most demanding in cultivation, succumbing to the slightest excess of moisture, particularly during the winter, and very prone to botrytis attack. *D. bryoides* is similarly temperamental, succumbing in particular to botrytis which causes rapid 'die back'. The botrytis attacks the resting rosette 'eyes' as well as the stem, especially during mild damp periods during the winter. *D. afghanica* also has a similar glandular habit and is equally vulnerable to fungal attacks – both species can be included amongst the group of 'the most difficult to cultivate' and are for that reason only found in the collections of highly skilful growers.

This is a great pity because, in other respects, *D. bryoides* is one of the most highly desirable species, neat and compact, slow-growing, pretty and floriferous. Most of the larger plants (the very largest is 10 cm in diameter, but this is quite exceptional) survive for 4–5 years – the oldest known is 8 years old. However, the average 3–4 year old plants are about 5–6 cm in diameter.

Cuttings have proved the easiest method of increase, although Eric Watson reports growing several plants from his home-produced seed. Cuttings taken in late May and June can be expected, under ideal conditions, to give a success rate of 60–80 %. Young cuttings are rather difficult to get through the first two years, but become rather easier once a small cushion has built up.

At one time both pin- and thrum-eyed plants were in cultivation so that a small amount of seed was occasionally produced. However, the thrum-eyed plants appear now to have been lost to cultivation.

Dionysia bryoides therefore requires diligent care if it is to be grown successfully. It is not an easy species to cultivate, but as Henrik Zetterlund remarks – 'This species is rather easy until the day of death – impossible to grow to any size . . . '. Chris Norton adds – 'I've lost this one several time and it pays to have 'lots' of cuttings in reserve'.

All growers report that *D. bryoides* is a very floriferous species, even small cushions smothering themselves in bloom.

Harold Esselmont in Aberdeen raised a particularly fine plant with broad-petalled flowers. This clone was lost in Britain, but maintained and propagated in Göteborg by Henrik Zetterlund who gave to it the cultivar name 'Harold Esselmont'. Plants from this source are now represented in several British collections.

Most clones flower during March and April in cultivation.

30 *Dionysia zagrica*

Dionysia zagrica Grey-Wilson in Kew Bull. 29 (4): 691, fig. 2, (1974).
Type: W Iran, Kuh-i-Sehquta, 36 miles (58 km) north of Pataweh (= Patureh), *Hewer* 2023 (holotype K; isotypes GB, W).

Description. *Plants* forming dense cushions (to 15 cm diameter in the wild), efarinose. *Stems* columnar, short-branched, becoming woody eventually, densely covered in overlapping dead marcescent leaf remains. *Leaves* oblong to oblong-

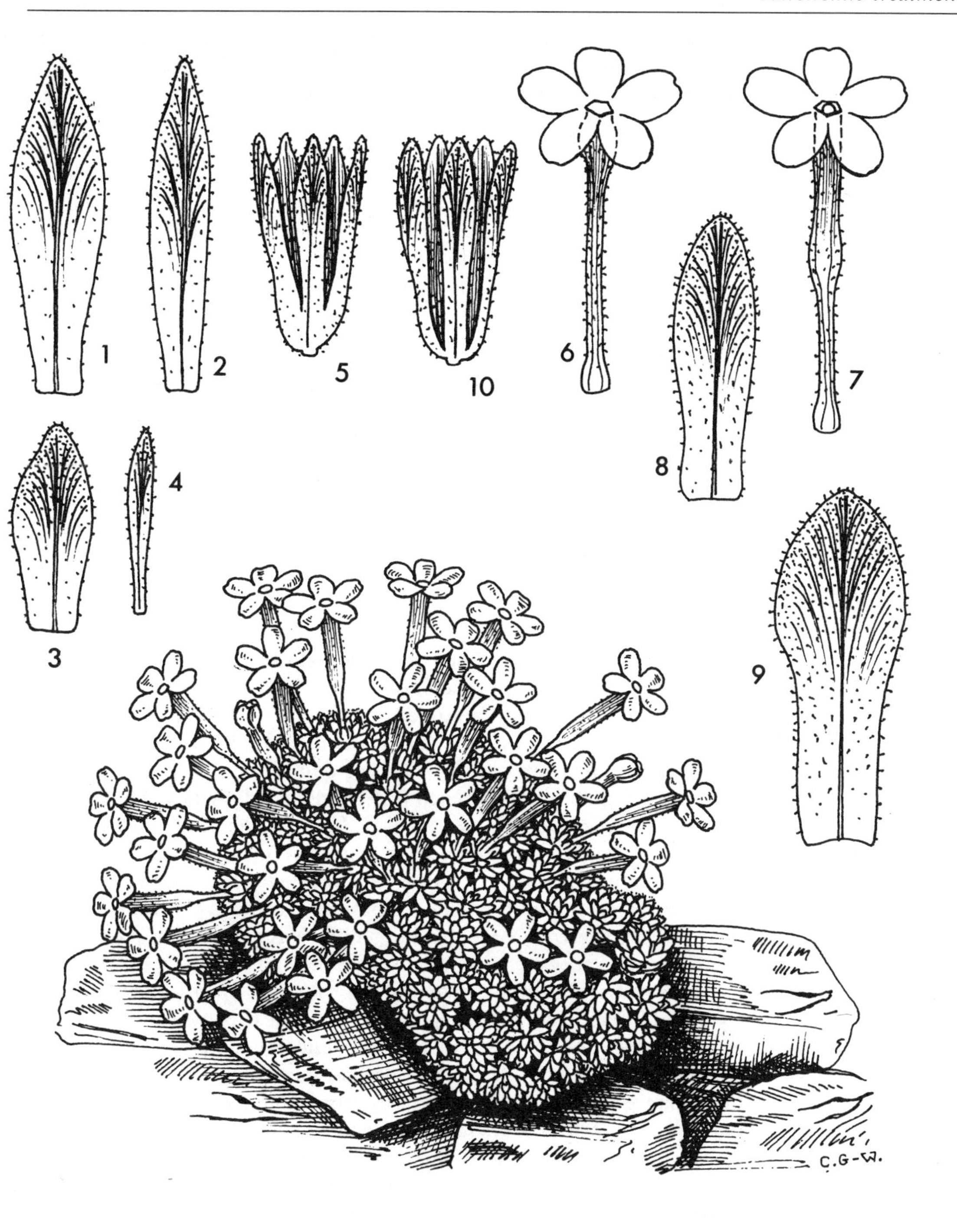

Dionysia zagrica, × 2. 1–3 leaves, × 12; 4 bracts, × 12; 5 calyx, × 9; 6 thrum-eyed corolla, × 4½; 7 pin-eyed corolla, × 4½. *D. sarvestanica*. 8–9 leaves, × 12; 10 calyx, × 9.

spathulate or obovate, 3–4.5 mm long, 0.75–1.4 mm wide, the apex obtuse to subacute, the margin entire, more or less densely covered in minute capitate glands; veins distinct and flabellate, especially below in the upper non-hyaline half. *Flowers* solitary, sessile, heterostylous. *Bracts* 1, oblong-linear, 2.5–3 mm long, glandular like the leaves. *Calyx* tubular-campanulate, c. 4 mm long, divided for three quarters of its length into oblong to narrow-spathulate, lobes, minutely glandular, flabellate-veined in the upper half. *Corolla* yellow, finely glandular outside; tube 10–14 mm long; limb 5–7 mm diameter, divided into suborbicular, obtuse or sometimes slightly retuse lobes. *Anthers* c. 1.25 mm long, inserted in the throat or in the middle of the corolla-tube. *Style* reaching the throat of the corolla-tube, occasionally slightly exserted, or not quite reaching the middle of the tube. *Capsule* containing 3–4 seeds (Plate 35).

DISTRIBUTION. W. Iran: Kuh-i-Sehquta just W of the Zagros Mountains, N of Pataweh; altitude 2400–2850 m.

HABITAT. Growing on north-facing limestone cliffs or exposed limestone slabs, in sun or partial shade. Flowering in the wild in April.

Dionysia zagrica was discovered in 1973 by Richard Hewer, who was accompanying his father Professor Tom Hewer on a botanical expedition to W Iran. The species was found on Kuh-i-Sehquta, an outlier of the Zagros Range which runs down the west side of Iran.

Plants form hard tight cushions which are a characteristic deep green colour. The calyces are rather larger than the other yellow-flowered species in subsection *Bryomorphae* and poke out conspicuously from the leaf-rosettes (though the flowers are sessile), in the dried state at least. The veins on the leaves and calyces, in the upper non-hyaline part, are raised and fan-like (flabellate), especially underneath; a character shared with *D. lamingtonii*.

Dionysia zagrica must be associated with the other yellow-flowered members of subsection *Bryomorphae*, *D. haussknechtii*, *D. lamingtonii*, *D. michauxii* and the recently described *D. sarvestanica*. It is readily distinguished from these species by the complete absence of articulated hairs on the leaves and calyces and by the dense covering of minute capitate glands. In this respect *D. zagrica* comes close to the violet-flowered *D. bryoides*, but it differs also in the almost entire, not deeply notched corolla-lobes. *D. sarvestanica* comes very close to *D. zagrica* in general characteristics, but differs primarily in the more open leaf-rosettes, the relatively broad leaves in which the upper half of the leaf-lamina is reflexed, the larger more leaf-like bracts, the calyx which is split to the base (rather than for two thirds of its length) and the shorter corolla-tube.

At first glance *D. zagrica* might be easily mistaken for *D. tapetodes* which is not represented in W Iran – *D. tapetodes* and its allies can be considered an eastern extension of subsection *Bryomorphae* (see p. 47), the glabrous corolla and the less deeply divided calyx distinguish *D. tapetodes*.

Dionysia zagrica is not in cultivation, although it is certainly a fine species and would make a valuable addition to live collections.

56 Darrah Belcheragh, NW Afghanistan, near Darrah Zang, on whose numerous N- and NW-facing limestone cliffs can be found *D. microphylla*. Photo. C. Grey-Wilson.

57 *Dionysia microphylla*. Photo. E. G. Watson.

58 *Dionysia microphylla*. Photo. E. G. Watson.

31 *Dionysia sarvestanica*

Dionysia sarvestanica Jamzad & Grey-Wilson in Kew Bulletin 44 (1): 124 (1989). Type: Iran, Fars, Post-i-Chenar, SE of Sarvestan, June, 1983, *Mozaffarian* 46754 (holotype TARI; isotype K).

DESCRIPTION. *Plants* forming dense caespitose grey-green cushions to 8 cm diameter; stems shortly branched, tightly packed with closely imbricating marcescent leaves, terminating in open leaf-rosettes. *Leaves* oblong-spathulate to narrow-obovate, 4–5 mm long, 1–1.5 mm wide, the apex obtuse, the margin entire, with distinct flabellate venation in the dried state, especially beneath in the upper non-hyaline part, covered on both surfaces by dense minute, sessile or short-stipitate glands, efarinose. *Flowers* solitary, sessile, heterostylous – only the long-style form is known. *Bracts* solitary, oblanceolate-rhombic, c. 4 mm long, leaf-like, glandular like the leaves. *Calyx* tubular-campanulate, c. 4 mm long, divided to the base into oblanceolate-rhombic, entire lobes, glandular. *Corolla* bright yellow, glandular outside; tube 8–10 mm long; limb c. 6 mm diameter, the lobes suborbicular, entire. *Anthers* c. 1.5 mm long, inserted in the middle of the corolla-tube. *Style* reaching the throat of the corolla-tube. *Capsule* containing ? 4–6 seeds.

DISTRIBUTION. Iran; Fars Province, Post-i-Chenar region, 10 km SE of Sarvestan, between Shiraz and Fassa; altitude 1700–2200 m.

HABITAT. North-facing limestone cliffs. Flowering ? April–May.

Dionysia sarvestanica is the most recently described species of *Dionysia*, discovered in Iran in June 1983 by an Iranian botanist, Mozaffarian. This species comes closest to *D. zagrica* – see p. 138 for general discussion of the similarities and differences.

Although *D. sarvestanica* and *D. zagrica* are fairly readily separated on present evidence (both are known only from their type localities), further collections from the Zagros Mountains may reveal intermediates. In such an eventuality the two species may need to be merged – the earlier name is *D. zagrica*.

Neither species is in cultivation, but with their dense small cushions of tight leaf-rosettes and bright yellow flowers they would certainly be worth introducing if seed can one day be collected from the wild.

32 *Dionysia denticulata*

Dionysia denticulata Wendelbo in Årbok Univ. Berg., Mat.-Nat. Ser. 19: 11, fig. 4 (1964). Type: Afghanistan. Bamian, between Shahtu and Panjao, June 1962, *Hedge & Wendelbo* W 4876 (hototype BG; isotype E).

DESCRIPTION. *Plants* forming rather loose cushions (to 50 cm diameter in the wild); farinose or efarinose. *Stems* columnar, short-branched, becoming woody below, above covered by closely overlapping dead leaves and leaf remains. *Leaves* elliptical to oblong or spathulate, 3.5–7 mm long, 0.75–2.5 mm wide, the apex acute to subobtuse, the margin denticulate in the upper half to subentire, covered on both surfaces with minute capitate glands; veins prominent, flabellate in the upper non-hyaline half. *Flowers* solitary, subsessile. *Bracts* 2, linear-lanceolate to elliptical, entire, 2–3 mm long. *Calyx* narrow-campanulate, 3–4 mm long, divided to about half-way into lanceolate-triangular, shortly mucronate, lobes, glabrous; limb 6.5–8 mm diameter, the lobes obovate to ± elliptical, distinctly emarginate. *Anthers* c. 1 mm long, inserted in the throat or just below the middle of the corolla-tube. *Style* reaching one third or three quarters the way along the corolla-tube. *Capsule* containing 7–8 seeds (Plates 36–7).

DISTRIBUTION. C Afghanistan: Bamian and Ghazni Provinces; vicinity of Behsud, NW of Ghazni, Malestan, Miradana. Okak, Panjao, San-i-Masha and Shahtu.

HABITAT. Shaded and semi-shaded sloping limestone rocks and cliffs, especially in narrow cliff crevices; altitude 2700–3300 m. Flowering in the wild from mid-March to May.

Dionysia denticulata is relatively rare in cultivation. Seed came from Afghanistan from two main sources – Per Wendelbo (1962) and Grey-Wilson & Hewer 1971 (GW/H 674). In the wild *D. denticulata* is extremely variable but nearly always very floriferous, like its close cousin *D. tapetodes*.

In cultivation *D. denticulata* tends to be short-lived – a 4 year old specimen would be quite exceptional. A 2 year old plant will reach 5–6 cm in diameter, a 3 year plant about 10 cm. Plants form rather lax, domed cushions.

Plants frequently flower in mid-winter (December-January), but sometimes also in the late summer. This early flowering habit would appear to leave plants open to botrytis attack. Most growers report that is is often difficult to keep plants healthy. Jack Wilkinson relates that – '... after flowering, 50% of first year plants will die after flowering the following year ... easy to grow at first, but hard to keep for 3 years' and Brian Burrow – 'flowers at Christmas and then usually dies. Only a small proportion survive'. Chris Norton tells the same story– 'I lose this plant with unfailing regularity – as soon as it flowers it dies – often difficult to find non-flowering rosettes for cuttings'.

Fortunately this short-lived plant is easy to propagate from cuttings. Henrik Zetterlund tells me that it is – 'Dead easy, May, 100%'. However, one can normally expect a success rate of 80–90%. This is perhaps just as well for had *D. denticulata* proved difficult to propagate then it may well have vanished from cultivation shortly after its introduction.

Flowers of cultivated plants are noticeably larger than those of *D. tapetodes* (which reacts quite differently in cultivation, see p. 37) and the petal-lobes are deeply notched. Those of *D. tapetodes* are entire or only slightly notched. The forms of *D. denticulata* in cultivation are all farinose.

Despite its short-lived characteristic *D. denticulata* is an attractive plant worth persevering with in cultivation.

For a comparison with the closely related *D. tapetodes* see p. 144.

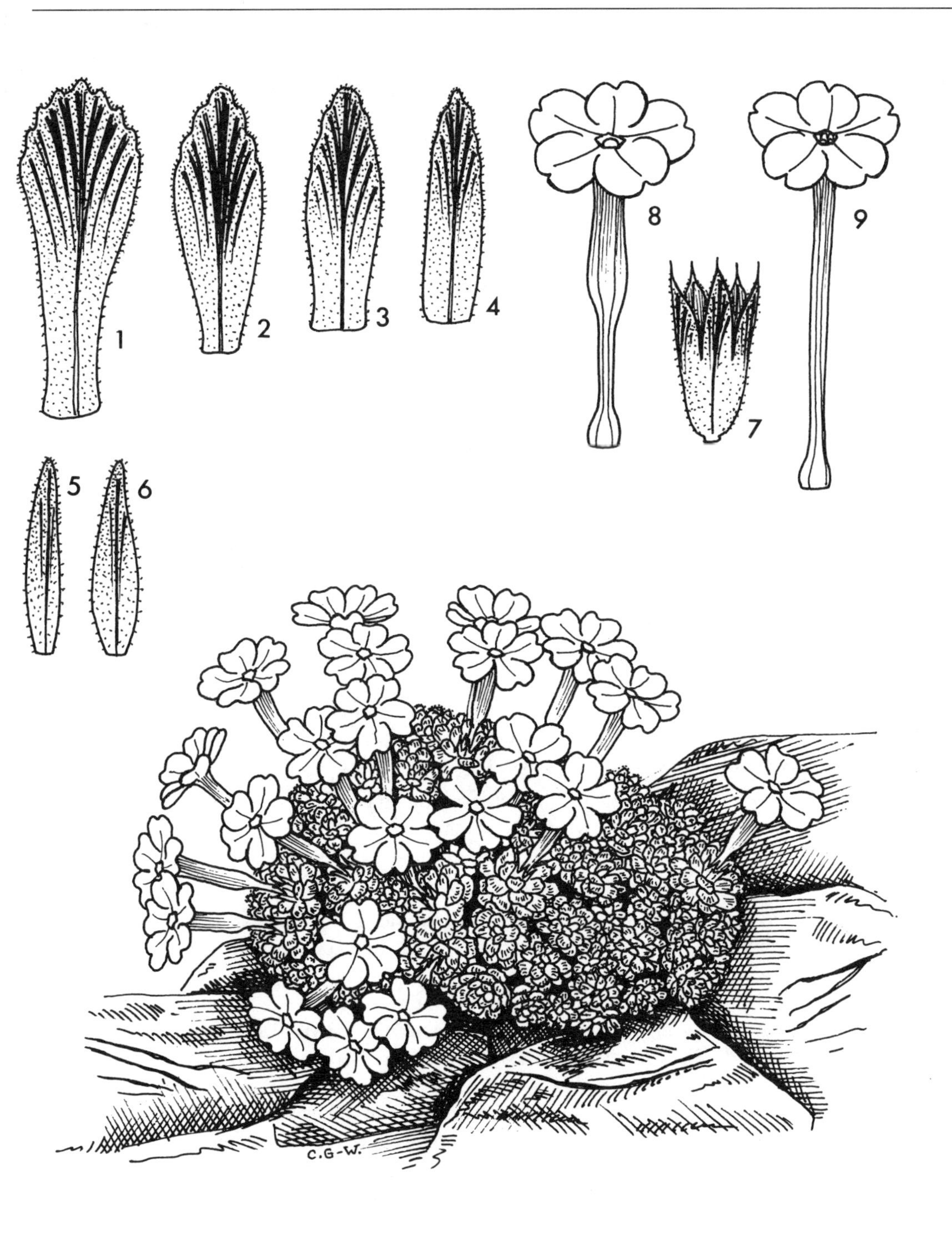

Dionysia denticulata, × 2. 1–4 leaves, × 12; 5–6 bracts, × 12; 7 calyx, × 6; 8 pin-eyed corolla, × 3; 9 thrum-eyed corolla, × 3.

33 *Dionysia tapetodes*

Dionysia tapetodes Bunge in Bull. Acad. Imp. Sci. St.-Petersb. 16: 562 (1871). Type: NE Iran, above Derrud, between Nishapur and Meshed (= Masshad), Bunge s.n. (holotype LE; isotypes G, P)

Primula tapetodes (Bunge) Kuntze, Revisio Gen. Plant. 2: 400 (1891).

Dionysia trinervia Wendelbo in Årbok Univ. I. Bergen, Mat.-Nat. Ser. 19: 14 (1964). Type: Afghanistan; Orozgan, Kotal-Tachakmak, 3100 m, June 1960, *K. Lindberg* 938 (holotypus BG).

DESCRIPTION. *Densely caespitose perennial* to 50 cm diameter, though often less, the plants forming flat or domed cushions. *Stems* much branched, covered in closely imbricate marcescent leaves, these forming columns closely packed together. *Leaves* oblong to elliptical, obovate or spathulate, 2–4 mm long, 1–1.5 mm wide, the apex acute to obtuse, the margin entire to slightly crenulate in the upper half, with prominent flabellate venation, especially beneath, covered on both surfaces with minute capitate glands, densely to sparsely covered in white or yellowish woolly farina, or efarinose. *Flowers* solitary, sessile, heterostylous. *Bracts* 1–2, linear to lanceolate, 2.5–3.5 mm long, glandular. *Calyx* tubular-campanulate, 2.5–4 mm long, divided for a half to two-thirds of its length into lanceolate or ovate, acute, ± entire lobes, , glandular, farinose inside or efarinose. *Corolla* yellow, very occasionally with a brown spot at the base of each lobe, glabrous; tube 9–14 mm; limb 5–6 mm diameter, divided into oval to suborbicular, entire or slightly emarginate, lobes. *Anthers* c. 1 mm long, inserted in the throat or in the upper third of the corolla-tube. *Style* reaching just below the middle of the corolla-tube, or reaching the throat, occasionally very slightly exserted. *Capsule* with 1–3 seeds. (Plates 38–42).

DISTRIBUTION. Widely distributed from the S USSR (Kopet Dagh), NE Iran (Mts of Khorrasan Province) to Afghanistan (Mts of Faryäb, Ghorat, Bamian, Baghlan, Parwan, Wardak, Kabul, Kapisa, Nangrahar, Paktria, Laghman and Badakhshan Provinces); altitude 1000–3200 m.

HABITAT. Growing in crevices and on ledges of limestone or occasionally dolomite cliffs, in shaded or partially shaded places, sometimes on sloping rocks or amongst boulders, especially beneath cliff overhangs. Flowering from early March to late May, very occasionally later.

Dionysia tapetodes was first collected by Griffith in Afghanistan between 1839 and 1841. Since then it has been collected on numerous occasions and there exist in herbaria more sheets of this species than any other.

Wendelbo (1961) placed *D. tapetodes*, together with the closely related *D. kossinskyi*, in subsection *Bryomorphae* Wend. Later (1965) he described two new species, *D. denticulata* and *D. trinervia*, placing all 4 species in a new subsection, *Tapetodes* Wend. This subsection was distinguished by the hairless, but glandular, nature of leaves, bracts and calyces; by the distinctive raised veins on the leaves which form a fan-shape, particularly beneath; by the glabrous corollas and the usually entire corolla-lobes.

Subsection *Bryomorphae* is restricted entirely to W Iran, whereas subsection *Tapetodes* comes from NE Iran, S USSR and Afghanistan. The characters of

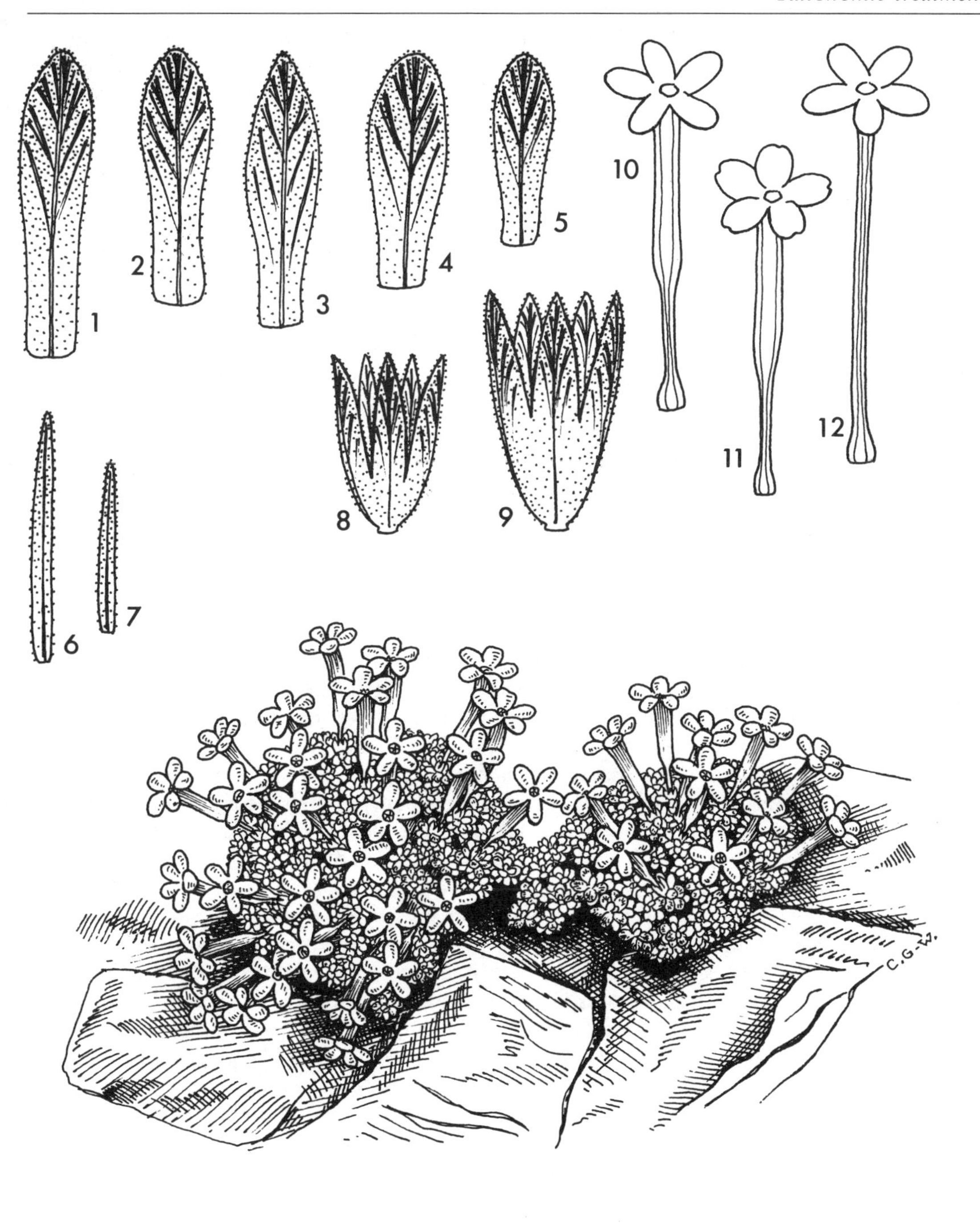

Dionysia tapetodes, × 2. 1–5 leaves, × 12; 6–7 bracts, × 12; 8–9 calyces, × 9; 10–11 pin-eyed corollas, × 4½; 12 thrum-eyed corolla, × 4½.

separation are only weakly defined. Both *D. bryoides* and *D. zagrica* in the *Bryomorphae* have glandualar (not hairy) leaves, bracts and calyces, as do the species of subsection *Tapetodes*. *D. lamingtonii* (which has hairy leaves) and *D. zagrica* have raised veins on the leaves, although these are more weakly defined than in *D. tapetodes* and its allies. *D. sawyeri*, like the species of subsection *Tapetodes*, has glabrous corollas. The species of *Bryomorphae* mostly have emarginate corolla-lobes, with the exception of the recently described *D. iranshahrii*. Subsection *Tapetodes* on the other hand possesses species with entire corolla-lobes, with the exception of *D. denticulata* in which they are distinctly emarginate. In *D. tapetodes* itself most plants possess entire corolla-lobes, but occasional forms are found with very slightly notched petals.

For these reasons it is not possible to uphold subsection *Tapetodes* and so in this revision I include them within subsection *Bryomorphae*. *D. tapetodes* and its allies represent an eastern extension of subsection *Bryomorphae*. Subsection *Scaposae* (of section *Anacamptophyllum*) has a similar, though more widely fragmented distribution.

Dionysia tapetodes has the widest distribution of any species of *Dionysia* in the wild, occuring in numerous localities from the Kopet Dagh in NE Iran and the neighbouring part of the USSR eastwards through the Hindu Kush of Afghanistan to the Pakistan frontier. As might be expected *D. tapetodes* is extremely variable over its range especially as regards the density and size of the cushions, the amount of farina, the size of the leaves and flowers and whether or not the corolla-lobes are entire or slightly notched. It has proved almost impossible to correlate all this variability so that it is not possible, nor indeed practical, to subdivide the species into subordinate categories; this includes forms with even more reduced leaves formerly assigned to *D. trinervia*.

Dionysia tapetodes has been introduced as seed a number of times in the past 30 years, indeed more forms of it exist in cultivation than for any other species in the genus. These can be summarized as follows: Paul Furse 1966 (PF 8964), Hedge, Wendelbo & Ekberg 1969, Grey-Wilson & Hewer 1971 (GW/H 780) and Hewer 1973 (H 1164). Plants under the numbers in brackets are often seen in collections, however, the source of many plants in cultivation has been lost – quite a number have in fact been raised from home-grown seed.

This species has been in continuous cultivation since 1958, the earliest seed being introduced by Mrs H. Priemer, although apparently no plants exist today from that source. In the *AGS Bulletin* for 1965 (33: 360) the plant mentioned under the name *D. caespitosa* is in fact *D. tapetodes*. The plant referred to was grown by Messrs Ingwersen and I quote Will Ingwersen's remarks – 'The plant was grown for 6 years in a compost of equal part tufa rubble, Sorbex peat, fine grit and loam, and was modestly fed with bone meal each spring. It is very slow-growing, adding only two or three of the tightly congested stems covered with minute, scale-like, grey-green leaves each year'.

The seed introductions of Paul and Polly Furse (1964 & 1966) brought in the species from a number of different localities. Jack Elliott wrote in 1968 – 'I have one plant of 1964 PF seed ... which has grown on steadily to about 4 in (10 cm) across ... I also have two or three plants from 1966 PF seed which seem to be less robust, but which have not yet flowered ... I have propagated the first one quite easily from cuttings in mist'. Peter Edwards wrote at the same time – 'another easy species to cultivate ... it seems it does not set seed, and after a free-flowering display it has a tendency to die'.

Since the 1960s, however, a lot more has been learnt about *D. tapetodes* in cultivation. Clones vary a good deal in vigour and floriferousness. Both farinose and efarinose forms are in cultivation. The most vigorous forms can reach up to 15 cm diameter in 4–5 years, 25 cm in 7–8 years.

Both pin- and thrum-eyed plants are in cultivation and several growers report that they have managed to get some seeds from their plants, sometimes by natural means, but also by careful hand-pollination.

However, the main way of ensuring a good supply of young plants is by taking cuttings regularly. *D. tapetodes* roots readily from single-rosette cuttings, sometimes giving a perfect 'take' of 100%, but 70–90% is more usual. Cuttings can be taken at any time from late May until September, rooting taking place in 4–6 weeks generally. Cuttings need to be guarded against botrytis attacks.

Botyrtis can also attack young and mature plants, especially during damp dull weather. Dieback of parts of the cushion, causing unsightly brown patches, may be initiated by botyrtis infection. Infected rosettes should be removed immediately lest they infect their neighbours. Treatment of the afflicted area with green sulphur powder can generally overcome the problem.

Most of the clones in cultivation are very floriferous, some being sweetly scented. Particularly vigorous forms such as H 1164 have proved shy-flowerers, though occasionally they are covered with bloom. The reason for this is unknown. Cultivated plants flower from January until late March. Most forms have bright yellow flowers. Some plants of GW/H 780 can be distinguished by the presence of a pale orange spot at the base of each petal. Similar spots occur also in *D. bornmuelleri* and *D. paradoxa*.

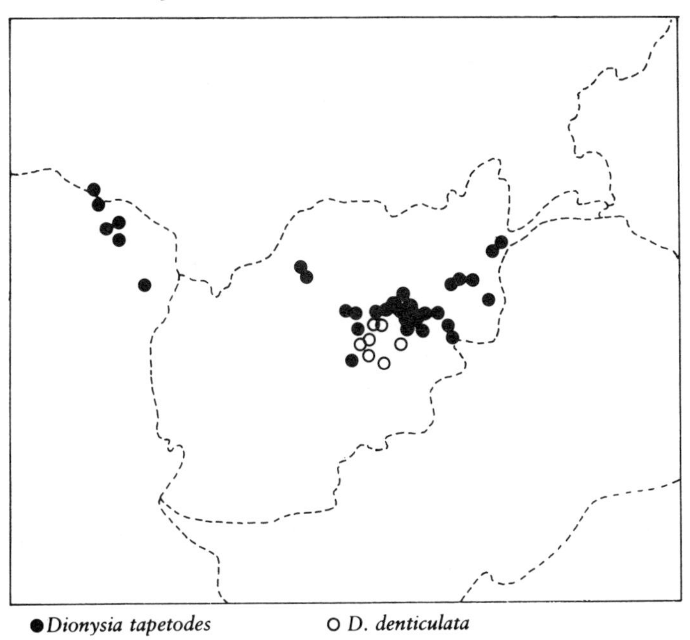

Map 6 Distribution of *Dionysia* subsect. *Bryomorphae* in Afghanistan and NE Iran.

34 *Dionysia kossinskyi*

Dionysia kossinskyi Czern. in Bull Jard. Bot. Princ. URRS 26: 116 (1927). Type: Iran, Khorrasan, Mt. Kisil-chisht, above Khorkei, *Czerniakovska* 375 (holotype LE – not seen).

DESCRIPTION. Very similar in general appearance and habit to *D. tapetodes*. *Leaves* ovate to obovate, 2–2.5 mm long, 1–1.5 mm wide, obtuse, margin entire, covered toward the base with short-stipitate glands, with raised flabellate ventation, especially beneath in the upper non-hyaline part. *Flowers* solitary, sessile, probably heterostylous – only the pin-eyed form is known. *Bracts* 2, lanceolate, 2–3.5 mm long, acute. *Calyx* tubular-campanulate, 2.5–3.5 mm long, divided halfway into lanceolate lobes, glandular. *Corolla* apparently brownish-violet, hairy outside; tube 9–11 mm long; limb c. 5–6 mm diameter, divided into obovate entire lobes. *Anthers* 1 mm long, inserted in the middle of the corolla-tube. *Style* of long-style corolla reaching almost to the throat. Capsule with 2–4 seeds.

DISTRIBUTION. Restricted to the Kopet Dagh on both the USSR and Iranian sides of the frontier – Mt. Kisil-chisht and Gaudan – not known elsewhere; altitude unknown but growing in high mountains.

HABITAT. Growing on cliffs, probably of limestone. Flowering during April, May and early June.

Very little is known of this rare species. It was described in 1927 from material collected in N Iran by Czerniakovska. Only one other collection exists – that of Lipsky collected in the Kopet Dagh on the Soviet side of the frontier that spans this mountain range.

Despite various attempts during the 1960s and 1970s by several people (notably Furse and later Hewer) to re-collect this species in its type locality, nothing has been seen of it since the earlier collections.

Dionysia kossinskyi is a puzzling species. It appears to be closely related to *D. tapetodes*, but further material is required before its exact status can be confirmed.

Dionysia kossinskyi differs from *D. tapetodes* in flower colour, by the corolla being hairy on the outside, by the leaf-indumentum and apparently in the size of the seed.

The flower colour is also somewhat perplexing as it is stated to be brownish-violet in the original description, but this is surely the colour of the dried floral remain. Live they may well be pink or violet.

35 *Dionysia lindbergii*

Dionysia lindbergii Wendelbo in Bot. Not. 112: 495, fig. 1 (1959). Type: Afghanistan, Darrah Zang, *Lindberg* 454 (holotype BG; isotype LD).

DESCRIPTION. *Plants* forming large soft rounded silvery-grey cushions (to 1 m diameter in the wild), efarinose. *Stems* thin, much-branched, covered in spreading, not closely overlapping, dead leaves and leaf remains. *Leaves* linear-

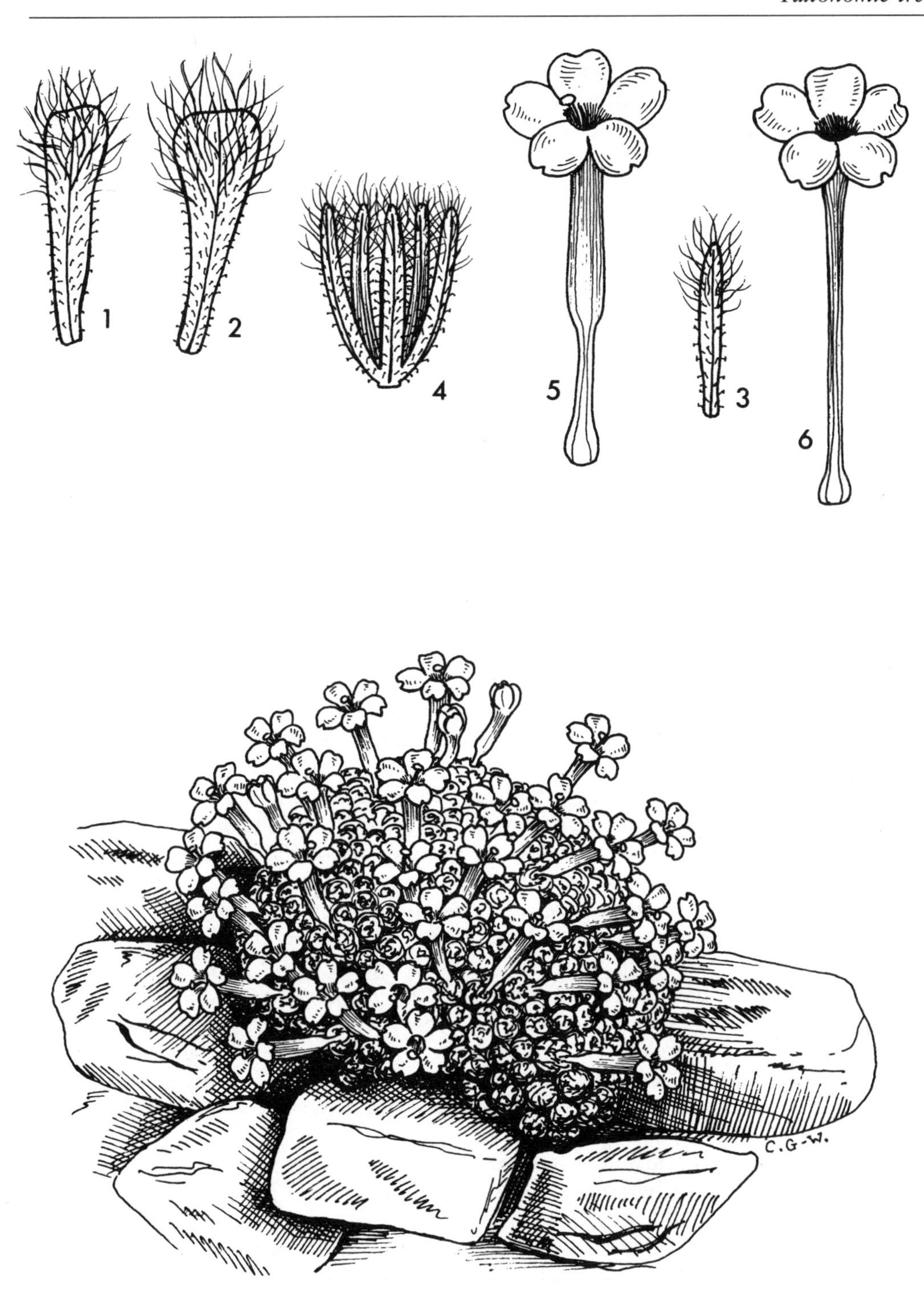

Dionysia lindbergii, × 2. 1–2 leaves, × 12; 3 bracts, × 4½; 4 calyx, × 4½; 5 pin-eyed corolla, × 4½; 6 thrum-eyed corolla, × 4½.

spathulate, 3.5–4 mm long, 0.4–0.8 mm wide, the apex obtuse to truncate or slightly emarginate, margin entire, covered all over with short-stipitate glands and towards the apex with long articulate hairs; veins rather obscure. *Flowers* solitary, subsessile, heterostylous. *Bract* 1, linear, 2.7 mm long, obtuse, hairy and glandular like the leaves. *Calyx* narrow-campanulate, c. 3 mm long, divided almost to the base into linear lobes, covered in short stipitate glands and at the lobe tips with long articulate hairs. *Corolla* violet, glabrous; tube 14 mm long, limb 9–10 mm diameter, divided into obcordate, deeply emarginate lobes. *Anthers* c. 1.5 mm long, inserted in the throat of the corolla. *Style* reaching one third the way up the corolla-tube. *Capsule* with 3 seeds (Plate 43).

DISTRIBUTION. NW Afghanistan; Maymana Province, Darrah Zang near Belcheragh – not known elsewhere; altitude c. 1400 m.

HABITAT. Growing in crevices on shaded limestone cliffs, often below overhangs or growing on the roof of cave entrances, generally on rather moist rocks; flowering during April and May.

Dionysia lindbergii is perhaps the most remarkable species in the genus. The large, soft, silvery-grey cushions, very hairy leaves and violet flowers are very distinct. Wendelbo considered the species so distinct that he granted it a subsection of its own, *Heterotrichae*, within section *Dionysia*.

As the species comes from N Afghanistan it is natural to link it with the violet-flowered members of section *Dionysiastrum*. Indeed, in the Darrah Zang gorge where *D. lindbergii* grows, it is found in close proximity to *D. microphylla, D. viscidula* and *D. afghanica*. The violet colour of the flowers of these species is, however, a tenuous link. All the members of section *Dionysiastrum* have very glandular leaves, bracts and calyces and the corollas are also glandular outside – none have articulated hairs. In contrast the leaves, bracts and calyces of *D. lindbergii* are glandular, but they are adorned with conspicuous long articulated hairs, especially towards the apex. Furthermore, the leaves are not closely imbricating as in section *Dionysiastrum* and the corollas are quite glabrous.

Dionysia lindbergii is therefore distinct in a number of characters. This interesting species seems to have no close allies within the genus, but appears to fit more comfortably in its own subsection within section *Dionysia*. There may be further, as yet undiscovered, species in the mountains of Afghanistan that may help to elucidate the true relationship of *D. lindbergii*.

The species was discovered in 1959 by K. Lindberg, a Swedish medical doctor and traveller. Plants form large soft cushions, often found in the wild growing upside down beneath overhangs or on the roof of caves, whence the seeds are presumably wafted by air currents. It can also be found on vertical walls.

Since it was first discovered *D. lindbergii* has been collected by Wendelbo & Hedge in 1962 and Grey-Wilson & Hewer in 1971. Seed from the latter collection (under GW/H 1304) produced several small plants which were raised to flowering size, but few survived more than three years and the species is no longer in cultivation. This is a great pity as *D. lindbergii* is a very beautiful species, the silvery-grey cushions forming domed hummocks supporting the small neat violet flowers. The soft, hairy nature of this species obviously presents a difficult challenge to the grower and it is likely to remain one of the most difficult of all to grow.

36 *Dionysia involucrata*

Dionysia involucrata Zapriagaev in Trudy Tadzh. bazy, botanika 2: 153, fig. 3, (1936). Type: USSR, Pamir-alai, Hissar (holotype LE).

DESCRIPTION. *Plants* forming rather dense deep green cushions (to 20 cm diameter in the wild). *Stems* short-branched, becoming woody below, above covered in closely overlapping dead leaves and leaf-remains. *Leaves* obovate to spathulate, 4–12 mm long, 2.5–6 mm wide, obtuse to subobtuse, generally with several (up to 5) small blunt teeth at the apex, densely covered on both surfaces with minute stipitate glands; veins distinctly raised and flabellate, on the upper surface especially. *Flowers* 3–5 in a scapose umbel, short-stalked, homostylous. *Scape* erect, 1.2–3 cm long, glandular. *Bracts* similar to the leaves, oval to suborbicular, 8–16 mm long, 5.5–17 mm wide, coarsely toothed, sometimes shallowly lobed, covered in minute stipitate glands. *Calyx* campanulate, 7–9 mm long, divided for three-quarters of its length into lanceolate, acute, entire lobes, glandular. *Corolla* violet-pink with a white 'eye' surrounded as the flowers age by a deepening band of violet-purple, stipitate-glandular outside, with powdery white farina towards the top of the corolla-tube and on the outside of the limb; tube 20–29 mm long; limb 9–14 mm diameter, divided into broad-obcordate, shallowly emarginate lobes. *Anthers* 2 mm, inserted two thirds the way up the corolla-tube. *Style* reaching the throat of the corolla-tube or shortly exserted. *Capsule* with 5–16 seeds (Plates 2, 45).

DISTRIBUTION. S USSR; Pamir-Alai, Hissarm–Khandar river gorge; altitude above 1500 m.

HABITAT. Growing in crevices of N- and NW-facing limestone cliffs; flowering during May and June.

Dionysia involucrata is endemic to the Pamir Mountains of the southern USSR bordering north-eastern Afghanistan. One other species, *D. hissarica*, is also endemic to the Pamirs although the two species do not grow in the same localities as far as it is known.

This species was first discovered in 1936 by Zapriagaev and for a number of years it was known solely from the dried type specimen. However, in 1975 seed was sent from the Leningrad Botanic Garden, (collected from wild plants) to Per Wendelbo at Göteborg University. From this source seed was sent to Britain and the species is now well established in cultivation.

Dionysia involucrata finds its closest ally in *D. hedgei*. Both species share the raised leaf-veins (characteristic of section *Dionysiastrum*), the finely glandular leaves, scapes and bracts, the scapose inflorescences and the violet-pink flowers. *D. involucrata* has leaves with a broader base and which are clearly toothed toward the apex. The bracts are larger than the leaves (of a similar size in *D. hedgei*) and coarsely toothed, sometimes almost lobed. The corolla-lobes are clearly notched, whereas they are entire in *D. hedgei*.

It is interesting to note that *D. involucrata* bears some farina (of the powdery type), but this is confined to the outside of the corolla, especially towards the top. In *D. hedgei* in contrast, there is often a good deal of farina on the leaves and inflorescences. In other members of subsection *Involucratae, D. freitagii, D.*

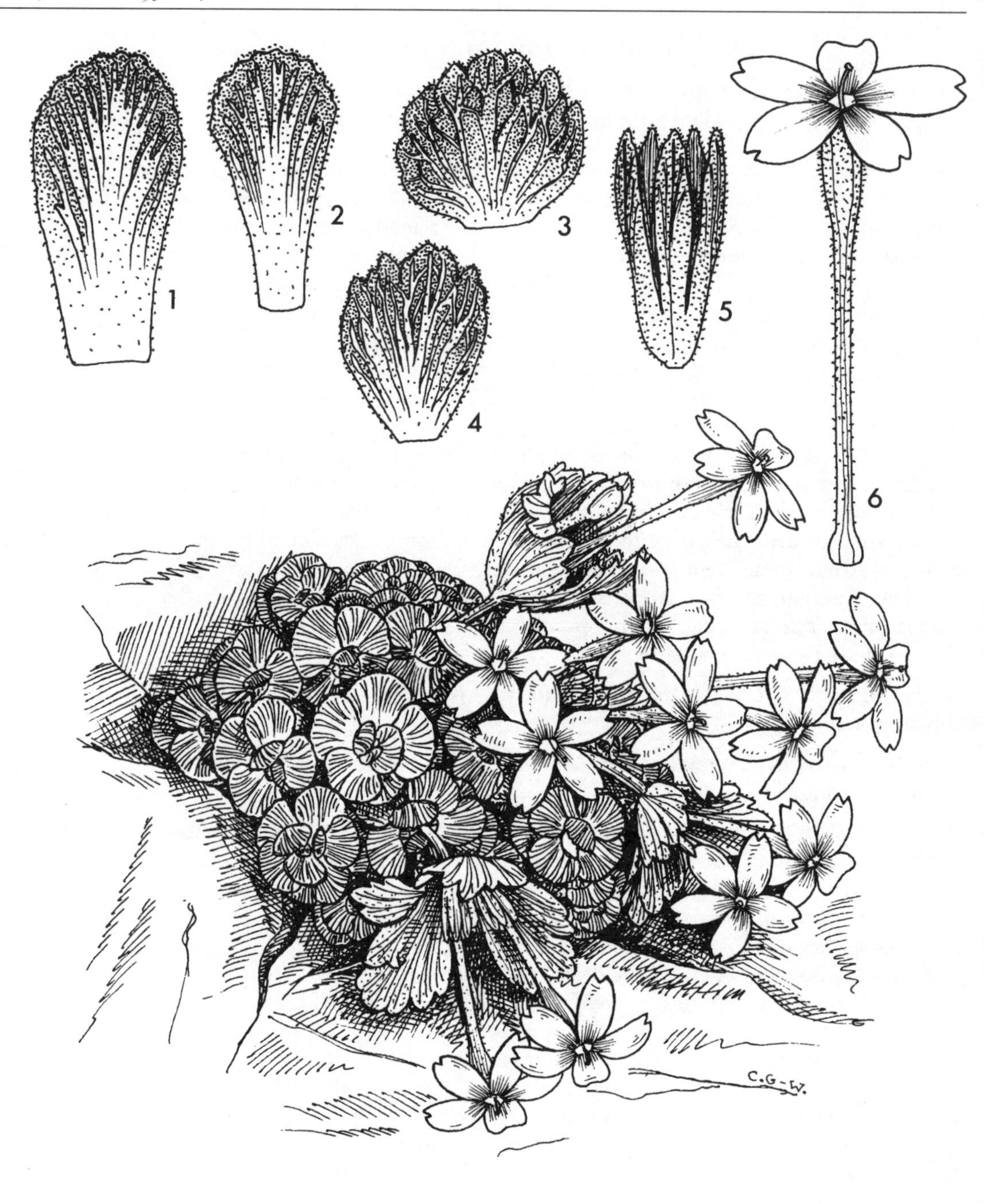

Dionysia involucrata, × 1½. 1–2 leaves, × 4½; 3–4 bracts, × 3; 5 calyx, × 4½; 6 corolla, × 3.

viscidula and *D. afghanica* (all scapeless species) farina is not present. *D. involucrata* appears to mark an intermediate stage, with farina very restricted on the plant.

Dionysia involucrata is unique in being the only species in the genus with homostylous flowers – flowers on all plants have similar flowers with stamens and styles set at the same level. As the flowers mature the style generally elongates pushing the stigma well beyond the stamens. This has its advantage in cultivated plants for seed is regularly produced making propagation less problematical. The scape can measure up to 30 mm long, but in cultivation it is often only 5–15 mm, the flowers appearing close to the cushion. In *D. hedgei* the flowers are generally borne well above the foliage. In both species the corolla 'eye' commences pale, sometimes whitish, but gradually it turns violet-pink, then red, as the corollas age. A similar characteristic occurs in a number of species of *Androsace* – *A. villosa* for instance.

Cultivated plants grow quite rapidly, achieving 15 cm diameter in 6–7 years. Specimens up to 18 cm diameter have been recorded, but many plants are only 8–12 cm diameter. At first the leaf-rosettes are relatively loose and large but as plants mature they become more compact, the rosettes held closely together.

Seeds generally germinate freely and the resultant seedlings are vigorous. Cuttings taken during May and June can produce a success rate of 60–80%, although young cuttings can be difficult to get through their first winter.

Seed is normally set naturally without any need to resort to time-consuming hand-pollination. Henrik Zetterlund reports that two albinos were produced from a batch of seedlings raised from 'home-saved seed' at Göteborg. This is an

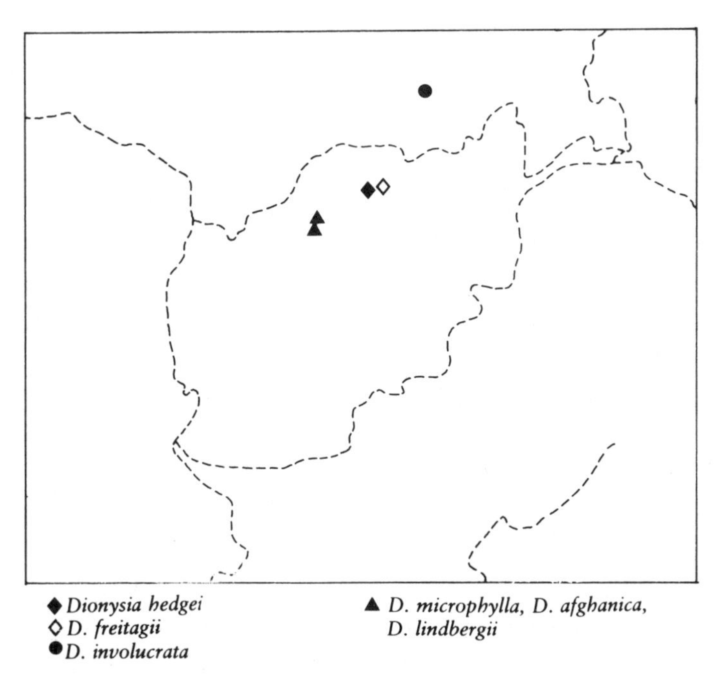

Map 7 Distribution of *Dionysia* section *Dionysiastrum*.

interesting occurrence for no other albino Dionysias have ever been recorded in cultivation. The white-flowered form has been distributed to a number of different growers.

Some growers find the plants not particularly long-lived in cultivation, but the stock can be readily maintained by regular sowing of seeds, or by rooting batches of cuttings annually. *D. involucrata* can be classed amongst the easier species to cultivate and well deserves a place in any collection.

37 *Dionysia hedgei*

Dionysia hedgei Wendelbo in Årbok Univ. I Bergin, Mat.-Naturv. Ser. 19: 18, figs 9–10, 16 (1963). Type: Afghanistan, Mazar-i-Sharif, Koh-i-Elburz, *Hedge & Wendelbo* W 3888 (holotype BG; isotype E).

DESCRIPTION. *Plants* forming rather lax, greyish-green cushions or hummocks (to 50 cm diameter in the wild), farinose. *Stems* becoming much thickened and woody below, above with spreading dead leaves and leaf-remains. *Leaves* oblanceolate to oblanceolate-spathulate, 5–9 mm long, 2–3.5 mm wide, thick, entire, acute or subacute, sometimes slightly apiculate, finely and minutely glandular-puberulous, above with whitish farina; veins indistinct beneath but prominent and raised above, generally with the farina in the grooves. *Flowers* short-stalked, up to 12 borne in a scapose umbel, heterostylous. *Scape* to 6.5 cm long, minutely glandular-pubescent. *Bracts* like the leaves, obovate to elliptical, 6–10 mm long, glandular and farinose. *Calyx* narrow tubular-campanulate, 7–8 mm long, divided for three-quarters of its length into linear-lanceolate, acute lobes, minutely glandular-puberulous inside and out and farinose inside. *Corolla* violet-purple with a darker centre, glandular-puberulous outside; tube 15–22 mm long; limb 12–14 mm diameter, divided into broad elliptic or obovate, very slightly emarginate lobes. *Anthers* c. 2 mm long, inserted in the throat or almost three-quarters the way up the corolla-tube. *Style* reaching two-thirds up the corolla-tube or exserted. *Capsule* with 4–9 seeds. (Plates 46–8)

DISTRIBUTION. N Afghanistan; Balkh Province, Koh-i-Elburz (Cheshma-i-Shafa, SW of Mazar-i-Sharif); altitude 500–1200 m.

HABITAT. Growing in crevices of limestone rocks, N- and NW-facing, on low cliffs or sloping rocks, often in full sun; flowering in April and May.

Dionysia hedgei was discovered in northern Afghanistan by Ian Hedge and Per Wendelbo on the 5 June 1962. Subsequent collections were made by Paul and Polly Furse in 1966, Professor Hewer in 1969 and Grey-Wilson and Hewer in 1971, and Wye College (Bob Gibbons), also in 1971.

The author, in the company of Tom Hewer had a chance to observe this striking species in flower during May 1971, close to the type locality. The Koh-i-Elburz (Alborz) mountains run from east to west to the south of the city of Mazar-i-Sharif not far from Afghanistan's border with the USSR. These mountains are much eroded and badly overgrazed; however, on the north side abrupt outcrops and cliffs of limestone harbour a number of interesting chasmophytic plants. Here can be found the remarkable little endemic *Scutellaria*

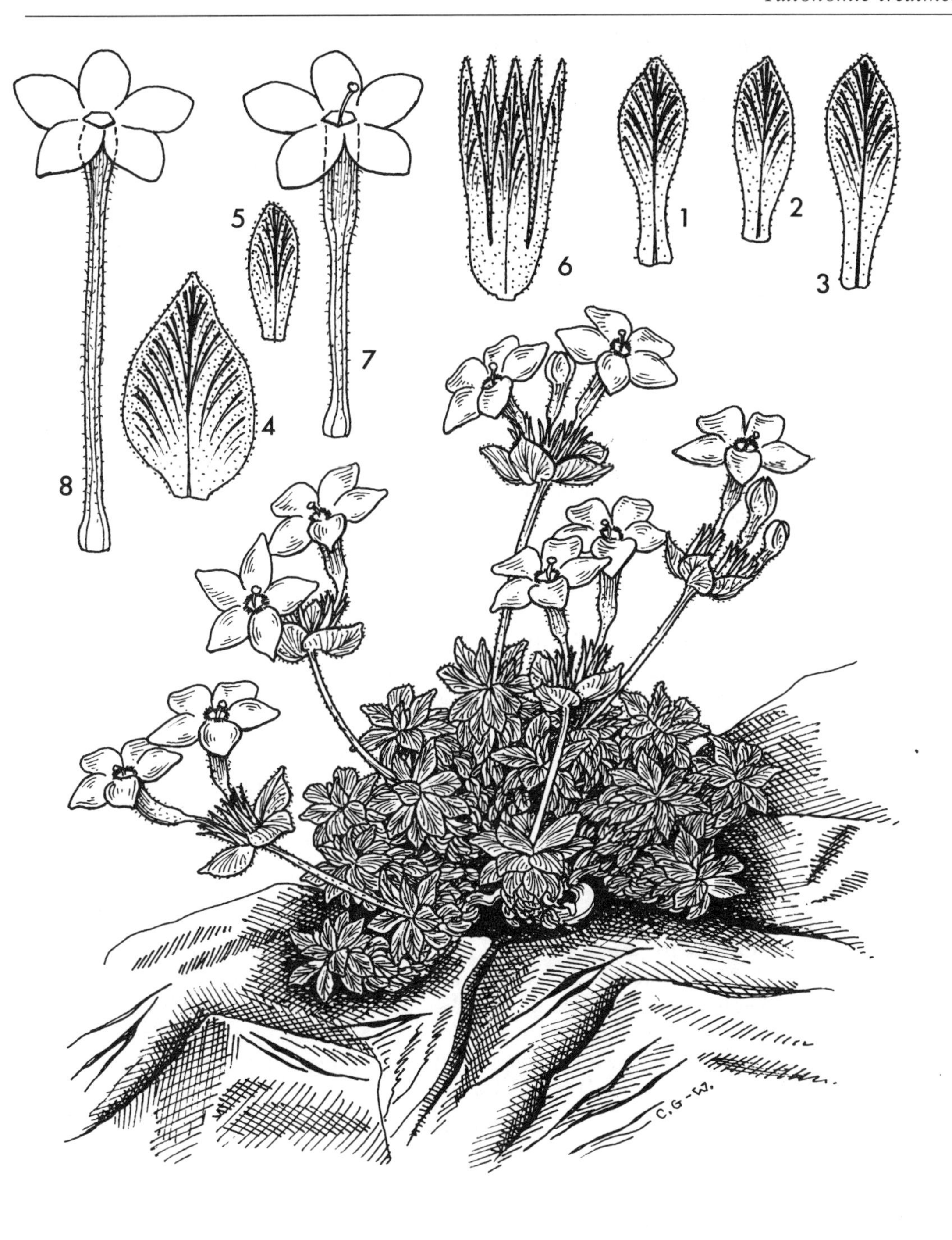

Dionysia hedgei, × 2. 1–3 leaves, × 4½; 4–5 bracts, × 4½; 6 calyx, × 4½; 7 pin-eyed corolla, × 3; 8 thrum-eyed corolla, × 3.

leptosiphon with its soft grey foliage and long-tubed palest pink flowers. Here also was found a new species of *Jurinea, J. perula-orientalis* C. Jeffrey. *Dionysia hedgei* was seen forming an extensive colony with many large old 'bushes' inhabiting vertical cliffs as well as sloping limestone rocks slabs. From a distance the flowers, which are a deep violet-purple, look like those of a rock Phlox, however, the long scapes held well above the foliage give this species a very distinctive look.

Dionysia hedgei is a substantial plant, perhaps the most shrubby of any *Dionysia*. There is little doubt that it comes closest to the Russian *D. involucrata*. The latter is confined to a small region of the W Pamir Mountains, the former to a limited area in the Koh-i-Elburz in N Afghanistan. Both species have similarly coloured flowers, well developed scapes, leafy bracts and leaves with raised veins. *D. involucrata* can be distinguished by its toothed leaves and bracts and notched corolla-lobes.

Dionysia hedgei would undoubtedly be one of the most desirable species to have in cultivation. Successive introductions of seed by Paul Furse, Grey-Wilson and Hewer and Bob Gibbons have produced no results – seed either failed to germinate or else subsequent seedlings soon perished.

38 *Dionysia freitagii*

Dionysia freitagii Wendelbo in Bot. Not. 123: 303, fig. 2A-F, (1970). Type: Afghanistan, Balkh, Ali Kuh, May 1969, *Hedge, Wendelbo & Ekberg* W 8497 (holotype GB; isotype E).

DESCRIPTION. *Plants* forming dense deep green cushions (to 40 cm diameter in the wild), efarinose. *Stem* woody below, above with closely overlapping dead leaves and leaf remains. *Leaves* variable, elliptical to subrhombic, 5–8 mm long, 2–3 mm wide, the apex subobtuse to acute, glandular-puberulous on both surfaces; veins prominent, flabellate, especially in the upper half above. *Flowers* solitary, subsessile, heterostylous. *Bracts* 2, linear-elliptic, c. 1.5 mm long, glandular-puberulous like the leaves. *Calyx* tubular-campanulate, 4.5–5 mm long, divided for a half its length into narrow-elliptic lobes, glandular-puberulous. *Corolla* violet to purple-violet, with a deeper 'eye', glandular-puberulous outside; tube 13–16 mm long; limb 8–12 mm diameter, the lobes broad-obovate, emarginate. *Anthers* 2 mm long, inserted in the throat or two-thirds the way up the corolla-tube. *Style* reaching the throat, or two-thirds up the corolla-tube. *Capsule* with 3–5 seeds (Plates 46, 49–50).

DISTRIBUTION. N Afghanistan: Balkh Province, Koh-i-Elburz (= Alborz), S of Mazar-i-Sharif (Ali Kuh, Ali Sher and close to Chahar Mahalla); altitude 900–1700 m.

HABITAT. N and NW facing limestone cliffs. Flowering in the wild in April and May.

Dionysia freitagii is one of the most spectacular of the violet-flowered Dionysias, forming neat rounded cushions in cultivation, smothered in the New Year by relatively large flowers. In the best forms these are a deep violet-pink with

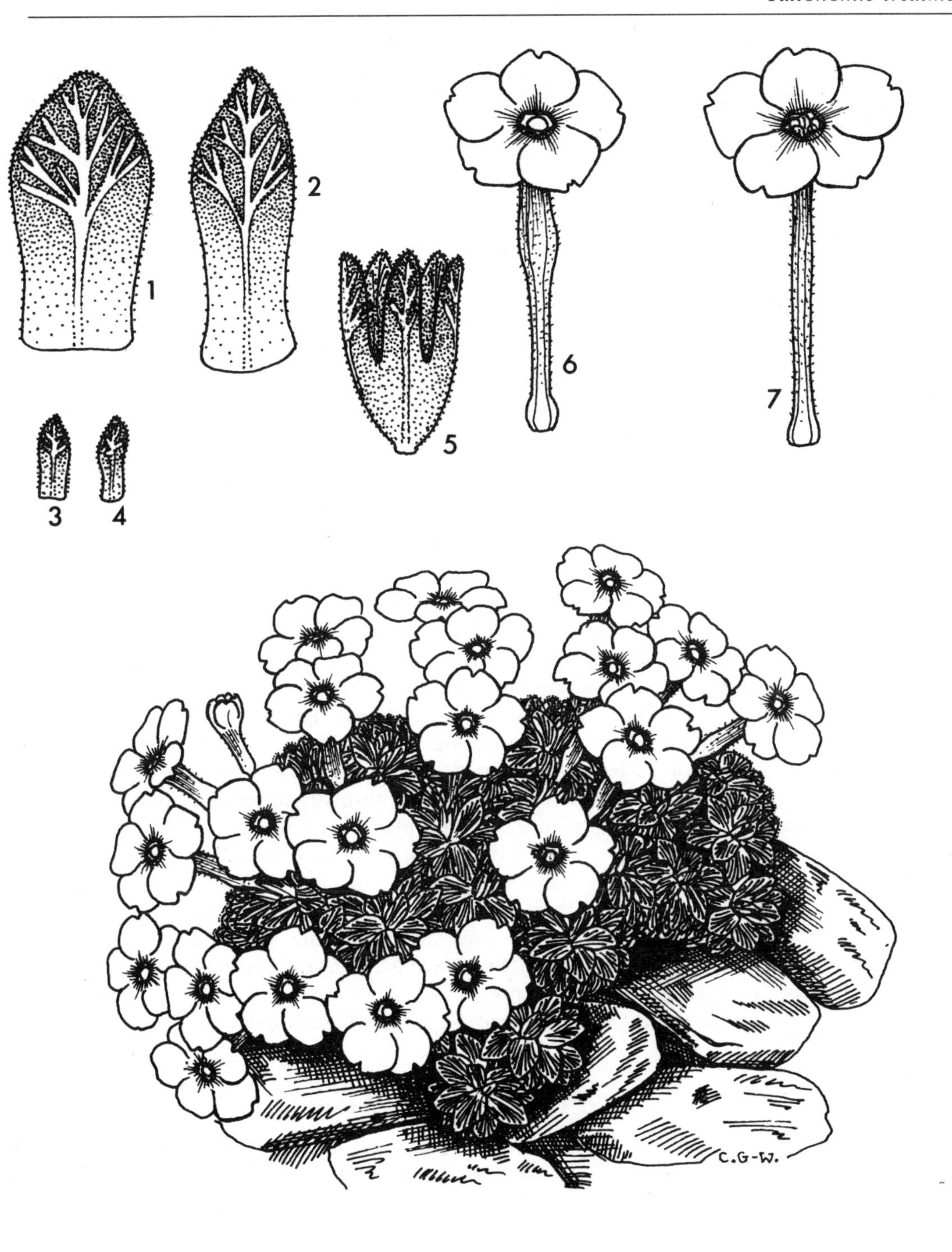

Dionysia freitagii, × 2. 1–2 leaves, × 9; 3–4 bracts, × 9; 5 calyx, × 6; 6 pin-eyed corolla, × 3; 7 thrum-eyed corolla, × 3.

a deeper 'eye' which surrounds a yellowish or whitish throat.

In my earlier account of the genus (1970) this species had not been formerly described and was included as *Dionysia* new species. It had been discovered only the previous year in northern Afghanistan by Hedge, Wendelbo and Ekberg. Seed was introduced by them, but to my knowledge no resulting plants were produced. In 1971 seed was introduced from two sources; Grey-Wilson & Hewer under GW/H 883 (generally incorrectly given as GW/H 8497) and Bob Gibbons *et al* collecting as a Wye College expedition, GFL 50827, which was introduced under the name *D. viscidula*; from these sources originated all the plants at present in cultivation.

In my earlier account I suggested that perhaps *D. freitagii* would be a 'difficult species in cultivation'. However, this has not been borne out and the species is generally considered relatively easy to cultivate, although not particularly easy to propagate. For this reason it is not very common in collections.

Plants form cushions up to about 15 cm diameter in 8–9 years, indeed they often make a plant of 1–2 cm diameter in their first year.

Cuttings are difficult to root and readily damp off. A success rate of 10% is good, 25% quite exceptional. Geoff Rollinson reports some success using Senagel instead of a grit- or pumice-based medium and this rooting gel certainly requires further trial.

The chief enemies of *D. freitagii*, as with the closely related *D. viscidula*, are greenfly and botrytis and plants need to be protected from each invasion.

Both pin- and thrum-eyed plants are in cultivation so that, with careful hand-pollination, it should be possible to produce a regular crop of seed. Eric Watson produces seed in this way and the resultant seedlings are vigorous.

Dionysia freitagii is a more vigorous species than its close cousin *D. viscidula*. Both species have rather sticky, deep green cushions. The flowers of *D. viscidula* are smaller and generally paler, normally appearing about a month later. *D. viscidula* is a far more difficult plant to cultivate.

In the wild *D. freitagii* inhabits shaded and semi-shaded limestone cliffs. In the single locality in the Koh-i-Elburz mountains, to the south-west of the city of Mazar-i-Sharif, where I have seen it, plants formed an extensive colony, some individuals about 40 cm in diameter, dense and rather hard. The cushions are packed with marcescent leaves, but towards the base the stems become thick and woody, especially in older plants.

Dionysia freitagii is a beautiful species which deserves to be more widely known.

39 *Dionysia viscidula*

Dionysia viscidula Wendelbo in Årbok Univ. I Bergen, Mat.-Nat. Ser. 19: 20, figs 11, 16 (1963). Type: Afghanistan, Darrah Zang, near Belcheragh, May 1962, *Hedge & Wendelbo* W 3722 (holotype BG; isotype E).

DESCRIPTION. *Plants* forming dense deep-green viscid cushions (to 30 cm diameter in the wild), efarinose. *Stems* short-branched, covered in closely overlapping dead leaves. *Leaves* elliptic-spathulate to oblong-oblanceolate, 5–8 mm long, 1.5–2.5 mm wide, subacute to subobtuse, entire, covered on both surfaces with minute stipitate-glands; veins distinctly raised (flabellate), especially on the upper surface. *Flowers* solitary, subsessile, heterostylous. *Bracts* 2, elliptic-spathulate, 2.5–3.5 mm long. *Calyx* campanulate, 3.5 mm long, divided for three-quarters of its length into elliptical lobes, stipitate-glandular. *Corolla* violet, generally with an 'eye', glandular-puberulous outside; tube 8–10 mm long; limb 7–10 mm diameter, divided into obovate, shallowly emarginate lobes. *Anthers* 1–1.5 mm long, inserted just above the centre or in the throat of the corolla-tube. *Style* reaching the throat, or one-third the way along the corolla-tube. *Capsule* with 5 seeds (Plates 51–2).

DISTRIBUTION. NW Afghanistan; Maymana Province, Darrah Zang near Belcheragh – not known elsewhere; altitude 1400 m.

HABITAT. Crevices in shaded and semi-shade limestone cliffs; flowering during April and May.

Dionysia viscidula was discovered by Ian Hedge and Per Wendelbo in Afghanistan in 1962. It was found growing on the soaring limestone cliffs of the great gorge at Darrah Zang in the north-west of the country. Per Wendelbo wrote in 1968 – '*D. viscidula* was growing on cliff walls much in the same way as *D. tapetodes*'.

The author, in the company of Professor T. F. Hewer travelled to Darrah Zang in 1971 and there found *D. viscidula* growing in mixed colonies with *D. tapetodes*, although the latter was by far the more common of the two species. The plants were seen and collected during August when the seeds were ripe and plants in cultivation today, under the number GW/H 1305, came from this source.

Dionysia viscidula finds its closest ally in *D. freitagii*, which occurs further to the east in Afghanistan. *D. viscidula* is an altogether slighter plant. It forms similar dark green sticky cushions, the leaves smaller and narrower with less prominently raised veins above. The flowers are small, mid violet-pink with a white, rather than deep violet 'eye'. In *D. freitagii* the leaves are arranged in more distinct whorls (glomerules) along the stem.

Dionysia viscidula is a very beautiful species. Unfortunately it is scarcely as easy in cultivation as *D. freitagii* and is rare in collections today. Plants grow slowly, attaining 6–9 cm diameter in 5–6 years. Only a single clone (pin-eyed) exists in cultivation and no seed has been produced. Unfortunately cuttings have proved very difficult to root with a success rate probably below 10%. At one time Eric Watson was the only person to grow this species and his diligence and skill as a propagator (as with the related *D. afghanica*) has kept the species in cultivation. Plants and cuttings are prone to attacks of botrytis and aphids and these need to be guarded against.

The species flowers in cultivation from February to April, coming into flower rather later than *D. freitagii*.

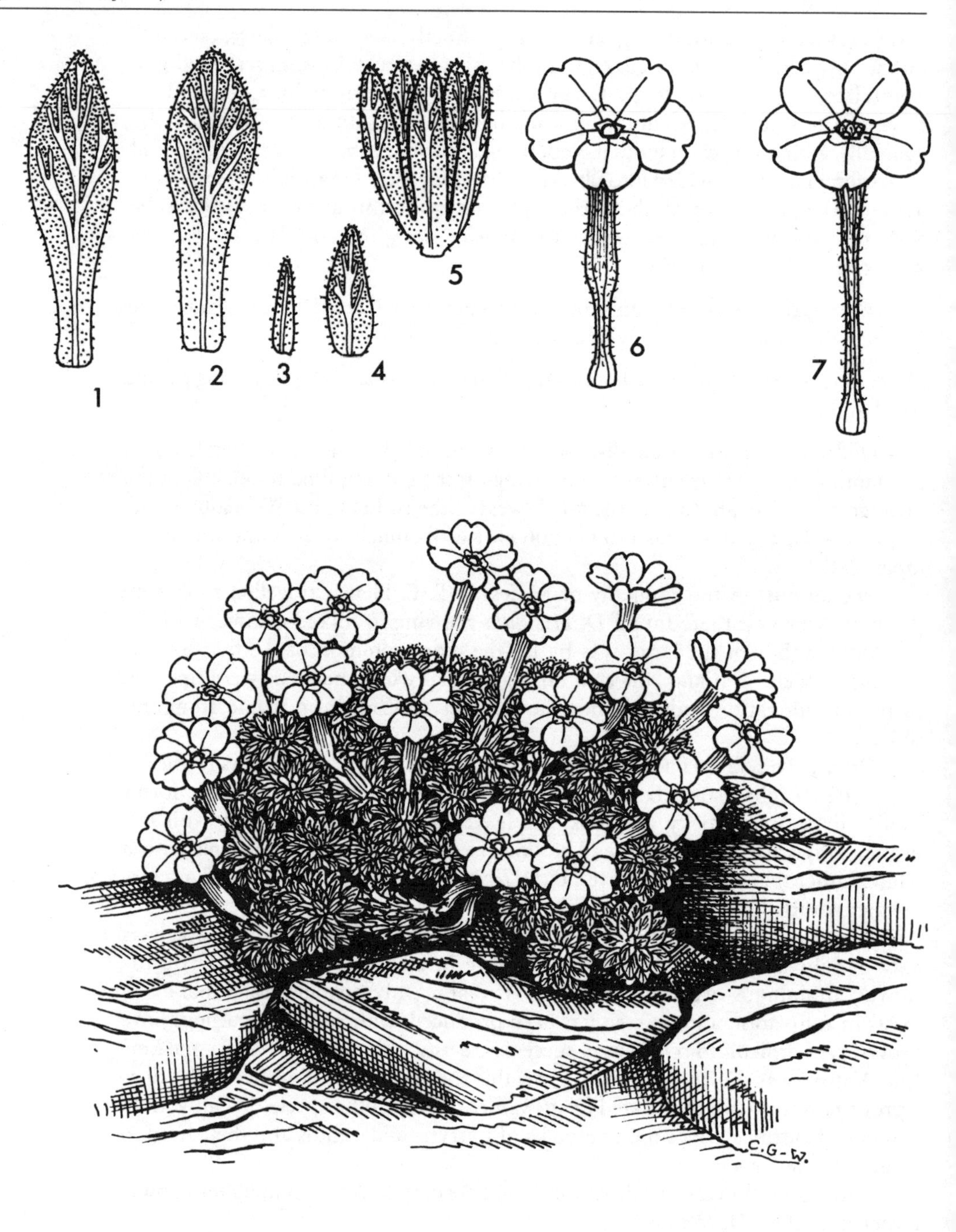

Dionysia viscidula, × 2. 1–2 leaves, × 6; 3–4 bracts, × 6; 5 calyx, × 9; 6 pin-eyed corolla, × 4½; 7 thrum-eyed corolla, × 4½.

40 *Dionysia afghanica*

Dionysia afghanica Grey-Wilson in Kew Bull. 29(1): 57 (1974). Type: Afghanistan, Darrah Zang, S of Belcheragh, July 1971, *Grey-Wilson & Hewer* 1308 (holotype K; isotypes E, GB, W).

DESCRIPTION. *Plants* forming dense, deep green cushions (to 20 cm diameter in the wild); efarinose. *Stems* columnar, short-branched, clothed by closely overlapping dead leaf-remains. *Leaves* oblanceolate to spathulate, 2.5–3.5 mm long, 1–1.25 mm wide, the apex subobtuse to somewhat emarginate, margin entire, covered on both surfaces with short glandular hairs; veins obscure. *Flowers* solitary, subsessile. *Bracts* solitary, linear-lanceolate, 2–2.5 mm long. *Calyx* campanulate, 2.5–3 mm long, divided for three quarters of its length into narrow-elliptical lobes, pubescent as the leaves. *Corolla* pale- to mid violet-pink with a darker 'eye'; tube 7–8 mm long, glandular-pubescent on the exterior; limb c. 7 mm diameter, the lobes obovate, emarginate. *Anthers* c. 1 mm long, inserted in the throat (thrum-eyed) or just below the middle of the corolla-tube. *Style* reaching just below the middle or three quarters the way along the corolla-tube. *Capsule* containing 1–2 seeds only (Plates 53–5).

DISTRIBUTION. NW Afghanistan: Faryäb Province, Darrah Zang, S of Belcheragh; altitude 1400 m.

HABITAT. Shaded limestone cliffs, on vertical walls and beneath overhangs. Flowering in the wild during ? April.

Dionysia afghanica is a distinct species probably finding its closest allies in *D. freitagii* and *D. viscidula*. However, the leaves lack a prominent flabellate (raised and fan-like) venation and the flowers have a dark 'eye' to the corollas. This is very distinct from the zones of deep violet or yellow in the centre of the corollas of the other two species – in these the colour zones are restricted to the limb of the corolla, whereas in *D. afghanica* it is confined to the throat.

Dionysia afghanica is only known from the Darrah Zang Gorge in NW Afghanistan where it was discovered by the author in the company of Professor Tom Hewer in July 1971. The gorge entrance is impressive with soaring cliffs tiered with bands of limestone. Parts of the bottom of the gorge are cultivated and there are some fine groves of walnut trees. Various paths run along the base of the cliffs, some traversing the occasional gentler slope. In the mouth of the gorge the semi-shaded cliffs harbour a rich assortment of chasmophytes including *Scutellaria ariana* and a cushion-forming *Viola, V. maymanica* Grey-Wilson. Dionysias abound on these cliffs, indeed more thickly so than the author has seen at any other locality. The commonest species by far in the gorge proved to be the widespread *D. tapetodes*, a pronouncedly farinose form. Amongst these, deep green sticky cushions proved to be the little known *D. viscidula*.

A little further along the gorge on the left side we clambered up the slopes to a wide ledge running by several large caves where Per Wendelbo had indicated we would find *D. lindbergii*. The species is almost exclusively confined to these overhanging rocks, especially the roofs of cave entrances. We did not see it anywhere else in the gorge.

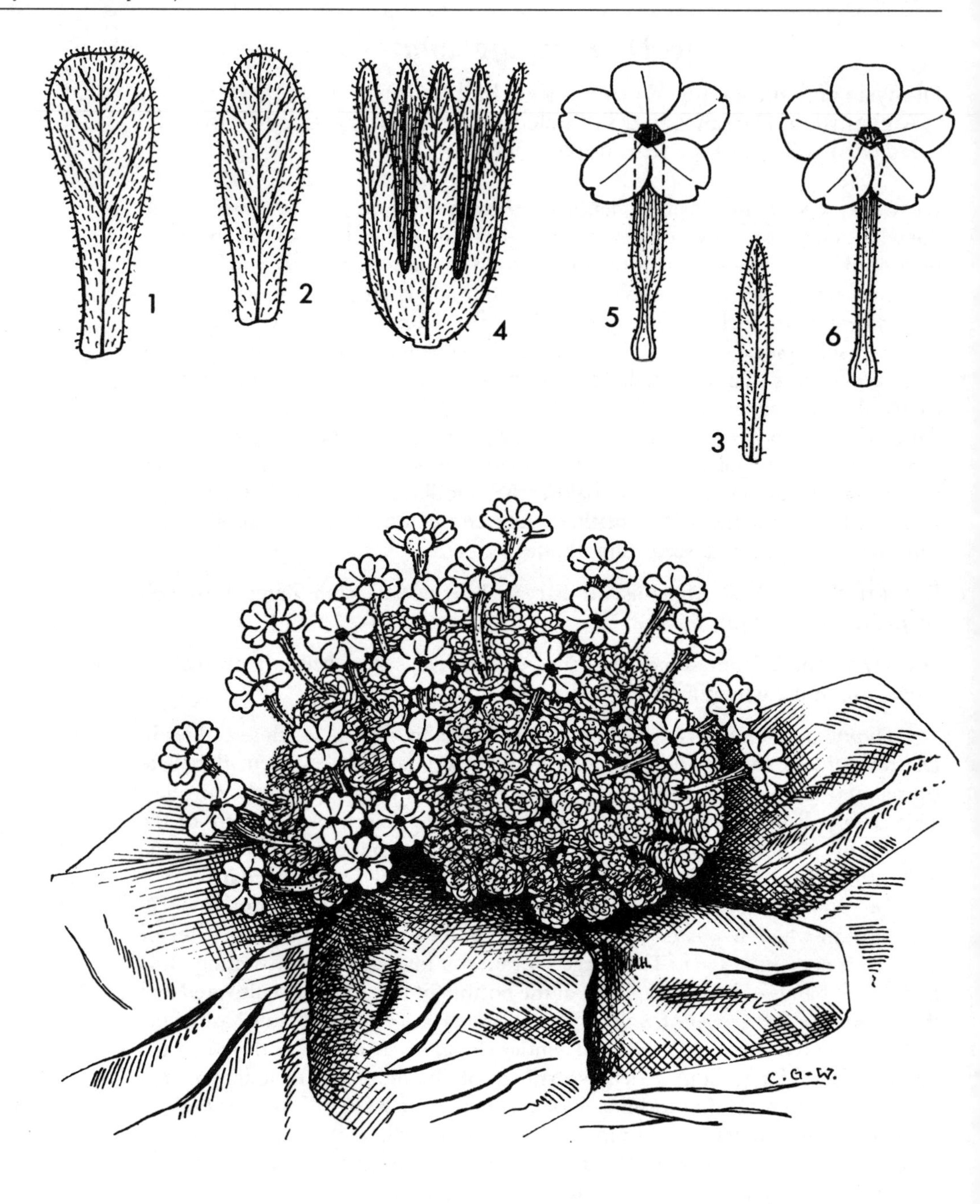

Dionysia afghanica, × 2. 1–2 leaves, × 15; 3 bracts, × 15; calyx, × 15; 5 pin-eyed corolla, × 4½; 6 thrum-eyed corolla, × 4½.

Continuing along the lower path on the left side of the gorge we passed beneath several large walnuts close to the cliff wall. Here a gentle overhang supported a large colony of Dionysias, all deep green with sticky cushions, the largest not more than 20 cm across. These were at first mistaken for *D. viscidula*, but a closer inspection soon told us that this was something quite distinct and exciting. Specimens were collected under GW/H 1308 and the species was described under the name *D. afghanica* on returning to Kew. As with most of the dense cushion-forming species the withered flower remains cling to the cushion so that by careful preparation (soaking gently and then boiling in water for about one minute) the details of the flowers can be analysed. Fragments of cushions were crushed and the detritus despatched to various growers upon return to Britain. Several seeds germinated initially but by 1980 only a single plant survived in cultivation. Eric Watson has related its history in cultivation as follows – 'I and one or two others managed to keep this going for a little while after its introduction, but it slowly died out. I failed to propagate it. Peter Edwards was the only person who kept it going and at the time of his death in 1979 he had two plants in his alpine house – one about 2 ins diameter and a smaller one about 1 in diameter. Jim Archibald had the big one and I had the small one. Unfortunately Jim lost his very quickly and I was left with the only one known to be in cultivation. I rooted one rosette in 1981 and during the same year the parent plant died. The single rosette flowered; the diameter of the flower was larger than the leaf-rosette. Once again the single rosette was the only plant in cultivation. It is a desperately slow growing species, but as soon as possible I took off a rosette and rooted it and over the years I've built up a little stock of plants that I have distributed to one or two enthusiasts ... I now have (1988) 12 plants from those in 3 in. pots to single rosettes. This species always flowers in January – often I have it in flowers on New Year's Day. The flowers last a long time, but soon fade from violet to a dirty white. The rosettes take a long time to root.'

Dionysia afghanica is not only one of the slowest growing species, but one of the earliest to come into flower each year. The oldest plants in cultivation are about 5 years old and the largest 6 cms diameter. The foliage is very sticky. Only the pin-eyed form exists in cultivation, so that there is virtually no chance of seed being produced and we must rely on skilled propagators to keep this difficult species in cultivation. The main danger to this charming plant is botrytis, especially during mild damp winter weather and they must be guarded carefully by severely regulating the water supply at that time of the year.

The clone in cultivation smothers itself in bloom, the flowers very neat and held very close to the cushion. Eric Watson remarks that – 'The cushions always cover themselves with flower – there is not even space for the petals to open entirely'. The flowers are easily recognised by their deep violet throat, quite unlike the closely related *D. viscidula* which has larger flowers zoned with white (occasionally pale yellow) around the throat.

41 *Dionysia microphylla*

Dionysia microphylla Wendelbo in Køie & Rech. fil. Symb. Afghan. IV, Biol. Skr. Dan. Vid. Selsk. 10(3): 68, fig. 26 (1958). Type: Afghanistan, precise locality unknown, but somewhere about 35°N 65°E, *Edelberg* 2313 (holotype C; isotype BG).

DESCRIPTION. *Plants* forming very dense grey-green cushions or tufts (to 40 cm diameter in the wild, though often smaller), farinose. *Stems* with very short branches, becoming thick and woody below, columnar, with numerous closely overlapping dead leaves. *Leaves* obovate to suborbicular, 1–1.5 mm long, 1–1.5 mm wide, rather thick, entire, apiculate, glabrous beneath, but above covered with minute capitate glands and with some yellowish farina; veins at apex somewhat raised above. *Flowers* 2–4 borne in an scapose umbel, or with up to 4 whorls, one above the other (superposed), each with up to 4 sessile flowers. *Scape* 2–26 mm long, glandular. *Bracts* suborbicular to ovate, 2.5–3.5 mm long, glandular and farinose like the leaves, *Calyx* tubular-campanulate, 4–5.5 mm long, divided for about two-thirds of its length into elliptical, subacute lobes, slightly and minutely hairy outside, glandular and farinose inside. *Corolla* mid- to deep violet, often with a darker purplish-violet centre and with a white 'eye', glandular-pubescent; tube 10–12 mm long; limb 8–9 mm diameter, divided into broad obcordate, shallowly emarginate lobes. *Anthers* 2.2 mm long, inserted in the throat or two-thirds the way up the corolla-tube. *Style* reaching one-third the way up the corolla-tube or somewhat exserted. *Capsule* with 3–4 seeds (Plates 56–8).

DISTRIBUTION. NW Afghanistan; Maymana Province, near Belcheragh (Darrah Abdallah and Darrah Zang); altitude 1200–1400 m.

HABITAT. Growing in crevices in limestone rocks, vertical cliffs or sloping rocks, in full sun or partial shade; flowering during April and May.

Dionysia microphylla is without doubt one of the most desirable species in cultivation. The hard cushions of small neat, rounded rosettes are beautifully formed and always attract attention. Add to these the pretty violet flowers, several borne on short scapes, and one has a most distinctive species.

Dionysia microphylla was discovered in 1953 by L. Edelberg, a Danish botanist, who was travelling in north Afghanistan. Unfortunately Edelberg did not record the locality of this find – furthermore the material which he collected was very scanty. However, in 1962 Ian Hedge and Per Wendelbo came upon the species close to Darrah Zang in north-western Afghanistan and full herbarium specimens were prepared.

Seed was introduced to Britain by Hedge and Wendelbo, later by Paul Furse (1966), Hedge, Wendelbo and Ekberg (1969) and by the author in the company of Tom Hewer (1971). Only seed from the latter collection, under GW/H 1302, resulted in flowering plants and all the plants currently in cultivation came from this source.

Dionysia microphylla is known from several localities near Belcheragh and in the gorge at Darrah Zang in north-west Afghanistan. There it inhabits vertical or inclined rock fissures or horizontal rock ledges, in limestone, often exposed to full sun, at least for part of the day. This may explain the very compact rather

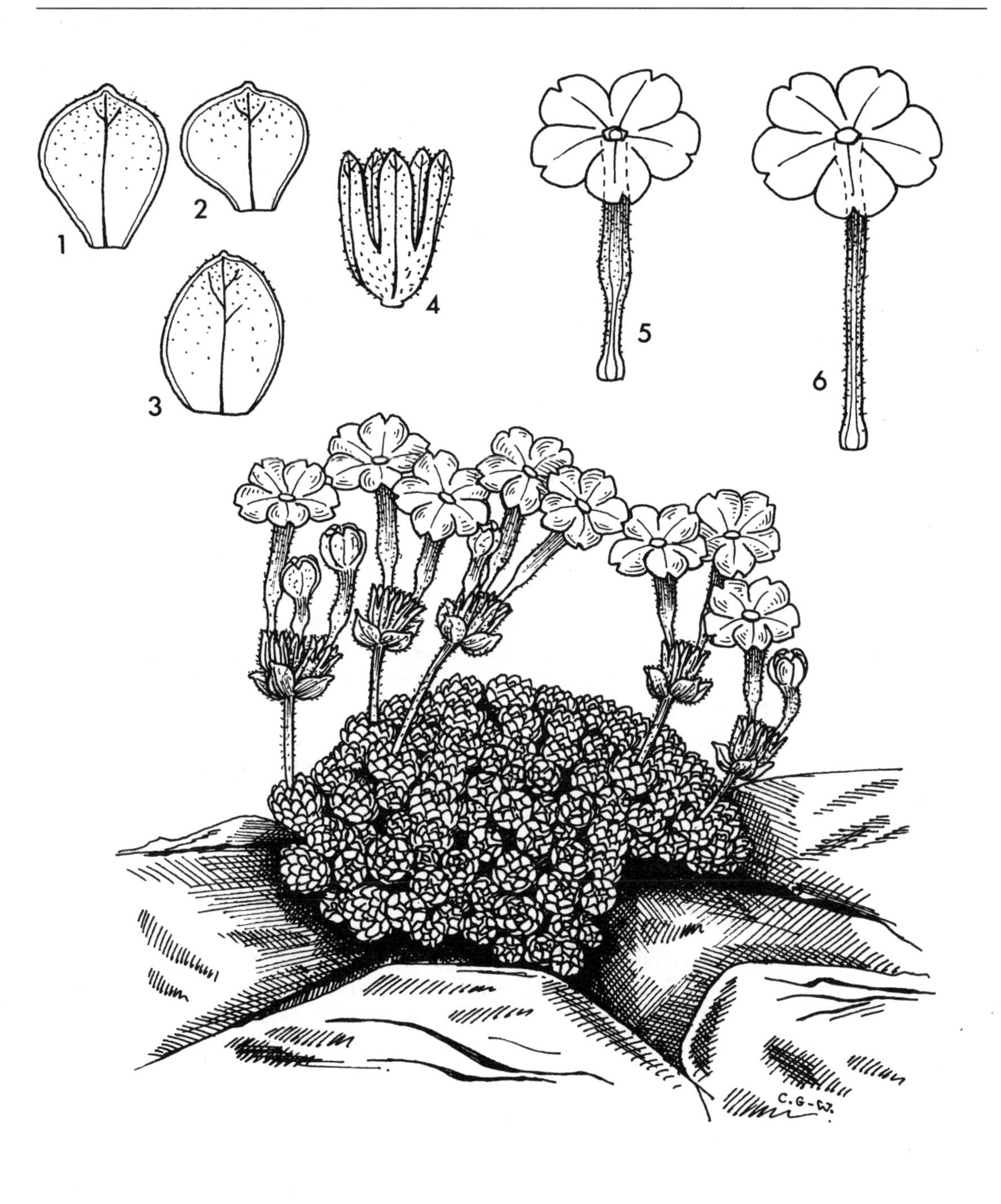

Dionysia microphylla, × 3. 1–2 leaves, × 18; 3 bracts, × 9; 4 calyx, × 4½; 5 pin-eyed corolla, × 4½; 6 thrum-eyed corolla, × 4½.

xerophytic look (the scale-like, thick, small closely overlapping leaves) of *D. microphylla*.

Plants grow rather slowly in cultivation, reaching 6–8 cm diameter in 4–5 years but as much as 15 cm in 7–10 years. They are relatively troublefree, but occasional attacks from both botrytis and aphids have been recorded. Most of the plants (if not all) appear to be pin-eyed, but despite this Brian Burrow reports obtaining a little seed by selfing plants through hand-pollination.

The rarity of this fine species in cultivation results not because it is particularly tricky to grow, but because it is very reluctant to root from cuttings. All growers report a very poor success rate, ranging from 0–5% only. This is a great shame as *D. microphylla* certainly deserves to be more widely cultivated. Henrik Zetterlund reports – 'Poor rooting – two cuttings (1%) have rooted over the years', but otherwise the species is 'rather troublefree'.

Plants flower during April and May in cultivation.

Dionysia hybrids

Dionysia hybrids are apparently very rare in the wild. This is not surprising as most of the species are isolated from one another. In the great gorge at Darrah Zang in north-western Afghanistan where no less than 5 species grow in close proximity (*D. afghanica, D. lindbergii, D. microphylla, D. tapetodes* and *D. viscidula*) one might perhaps expect to see a hybrid or two. However, no one has been to the gorge whilst they are in flower and it would be difficult to spot hybrids out of flower.

The only report of a hybrid *Dionysia* in the wild came from Professor Hewer who travelled in the Zagros Mountains of Iran during 1973. On the cliffs of the Kuh-i-Dinar near the town of Seesakht he came upon two species; the pink-flowered *D. bryoides* and the yellow-flowered *D. termeana*. On one small section of cliff several plants were found which were clearly hybrids with yellow flowers, the corolla-lobes edged with pink or red. Photographs and carefully prepared herbarium specimens backed up the find and these were later carefully analysed at Kew.

This interesting hybrid shows almost intermediate characters between the parent species. It is not known whether this hybrid is sterile or fertile.

In cultivation several hybrids were produced at Göteborg in Sweden. They are as follows:

Dionysia aretioides × *D. bornmuelleri*
Dionysia aretioides × *D. mira*
Dionysia aretioides × *D. teucrioides*

It is particularly interesting that of the various hybrids attempted, *D. aretioides* has proved a 'key species' in producing successful artificial hybrids. It is also interesting that all the species listed above belong to section *Anacamptophyllum; D. aretioides* fits into subsection *Revolutae*, the other three species to subsection *Scaposae*. This is of particular interest botanically as it clearly shows the close relationship of the species in these two subsections.

Dionysia aretioides × D. teucrioides. A particularly attractive hybrid that inherits the large flowers and easy cultivation of *D. aretioides*. Plants form large loose mounds, attaining 30 cm across in 5–6 years. The flowers are slightly smaller than *D. aretioides* but several are borne on each scape making plants extremely floriferous. The corolla is a deep yellow with a touch of orange (inherited from *D. teucrioides*).

Cuttings struck in June or July root readily and plants seem to be reasonably easy to 'grow on'.

Dionysia aretioides × D. mira. Both these species are easy in cultivation and the hybrid inherits this characteristic. The hybrid bears intermediate characteristics and is extremely vigorous forming a hemisphere 60–70 cm across in 5–6 years. Both pin- and thrum-eyed plants exist in cultivation. Henrik Zetterlund in Göteborg reports that this hybrid does very well planted on a tufa block. Propagation is easy from cuttings.

Dionysia aretioides × D. bornmuelleri. This is another extremely vigorous hybrid forming a mound up to 50 cm across in 7–8 years. In general characteristics the hybrids are intermediate between the parent species; several flowers are borne on each scape, they are quite large and flat, not as large as *D. aretioides*, but without the small curled corolla-lobes of *D. bornmuelleri*. Cuttings root readily during June and July and, like the previous species, plants thrive particularly well on a tufa block.

These three hybrids are all well worth cultivating and are attractive and floriferous plants. They deserve to be more widely grown and distributed. Other hybrids may occur in time though it is highly unlikely that hybrids between the three main sections of the genus will be easy to produce.

Appendix

Subsection *Afghanicae*, subsectio nova. Folia integra, dense imbricata, vix nervato, stipitato-glandulosa. Flos solitarius, sessilis, scapo obsoleto; corolla violacea, lobis emarginatis. Typus subsectionis: *D. afghanica* Grey-Wilson.

Bibliography

BALLS, E.K. (1934). Plant collecting in Persia. *Gardn. Chron.* ser. 3, 95: 44–45, 78–79, 346–347; 96: 40–41.

BOISSIER, E. (1846). *Diagnoses Plantarum Orientalium Novarum* 1(7). Lipsiae, Parisiis.

BOISSIER, E. (1879). Flora Orientalis. 4.

BOKHARI, M.H. & WENDELBO, P. (1976). Anatomy of Dionysia: Foliar sclereids. *Notes Roy. Bot. Gard. Edinb.* 131–141.

BORNMÜLLER, J. (1899). Drei neue Dionysien. *Bull. Herb. Boiss.* 7: 66–74.

BORNMÜLLER, J. (1903). Weitere Beiträge zur Gattung Dionysia. *Ibid.* ser. 2, 3: 590–595.

BORNMÜLLER, J. (1904). Dritter Beitrag. zur Kenntnis der Gattung Dionysia. *Ibid.* 4: 513–521.

BORNMÜLLER, J. (1905). Vierter Beitrag zur Kenntnis der Gattung Dionysia. *Ibid.* 5: 261–263.

BORNMÜLLER, J. (1911). Collectiones Straussianae novae. *Beih. Bot. Centralbl.* 28(2): 458–535.

BORNMÜLLER, J. (1915). Reliquiae Straussianae. *Ibid.* 33(2): 165–269.

BORNMÜLLER, J. (1937). Fünfter Beitrag zur Kenntnis der Gattung Dionysia. *Feddes, Repert. Sp. Nov.* 41: 179–180.

BRUNN, H.G. (1932). Cytological studies in *Primula. Symb. Bot. Upsal.* 1: 1–239.

BUNGE, A. (1871). Die Arten der Gattung Dionysia Fenzl. *Bull. Acad. Imp. Sci. St.-Petersb.* 16: 547–563.

CLAY, S. (1937). *The present-day rock garden.* London. 193–195.

CLUSIUS, C. (1605). *Exoticorum libri decem.* Antwerpiae.

CZERNIAKOVSKA, E. (1927). Dionysia kossinskyi. *Bull. Jard. Bot. Princ. URRS.* 26: 116.

DAVIS, P.H. (1947). On the rocks. *Bull. Alp. Gard. Soc.* 15: 37–53.

DAVIS, P.H. (1951). Cliff vegetation in the eastern Mediterranean. *Journ. Ecol.* 39: 63–93.

DECROCK, E. (1901). Anatomie des Primulacées. *Ann. Sci. Nat. Bot.* 13: 1–199.

DUBY, J. E. Primulaceae in DC., *Prodr.* 8: 33–74.

DYER, W. T. T. (1887). Note on the Botanical identification of Hamama. *Pharm. Journ. Trans.* 1887: 232.

FARRER, R. (1919). *The English rock-garden* 1. London.

FARVARGER, C. (1958). Contribution à l'ètude cytologique les generes Androsace et Gregoria. *Veröff Geob. Inst. Rübel* 33: 59–80.

FENZL, E. (1843). Plantarum generum et specierum novarum. Decas prima. *Flora* 26: 389–404.

GREY-WILSON, C. (1970). *Dionysia – the genus in the wild and in cultivation.* Alpine Garden Society.

GREY-WILSON, C. (1974). Some notes on the flora of Iran and Afghanistan. *Kew Bull.* 29(1): 19–81.

GREY-WILSON, C. (1974). Some notes on Iranian Dionysias (Primulaceae). *Ibid.* 29 (4): 687–694.

GREY-WILSON, C. (1976). *Dionysia, supplement and key.* Alpine Garden Society.

GREY-WILSON, C. (1989). A new species of *Dionysia* from southern Iran. *Kew Bull.* 44(1): 123–5.

GIUSEPPI, P. L. (1933). Some alpine plants of Persia. *Bull. Alp. Gard. Soc.* 2: 1–4.

GIUSEPPI, P. L. (1944). Some Iranian Dionysias. *Ibid.* 12: 85–86.

HOLMES, E. M. (1887). The botanical source of hamama. *Pharm. Journ. Trans.* 1887: 151–152.

HOLMES, E. M. (1887). The botanical identification of hamama. *Ibid* 252.

JAMZAD, Z. & GREY-WILSON C. (1989). A new species of Dionysia from Southern Iran. *Kew Bull.* 44(1): 124

JAUBERT, C. & SPACH, E. (1842–43). *Illustrations plantarum orientalium.* I Paris

KNUTH, R. (1905). Dionysia in Pax, F. & Knuth, R. Primulaceae. *Pflanzenreich* 4(237): 160–168. Leipzig.

LACE J. H. & HEMSLEY, W. B. (1891). A sketch of the vegetation of British Baluchistan with descriptions of new species. *Journ. Linn. Soc. (bot.)* 28: 288–327.

LEHMANN, J. G. (1817). *Monographia generis Primularum*. Lipsiae.

LIPSKY, V. I. (1900). Materialy dlia flory Srednei Azii. I. *Trudy Imp. St.-Petersb. bot. sada*. 18: 1–146.

LIPSKY, V. I. (1904). Materialy dlia flory Srednei Azii. II. *Ibid*. 23: 1–247.

MELCHIOR, H. (1943). Entwicklungsgeschichte der Primulaceen – Gattung Dionysia. *Mitt. Thür. Bot. Ver. N. F.* 50: 156–174.

PARSA, A. (1949). *Flore de l'Iran*. 4. Tehran.

PAX, F. (1889). Monographische Übersicht über die Arten der Gattung Primula. *Bot Jahrb*. 10: 75–241.

PAX, F. (1905). Primula in Pax, F. & Knuth, R. Primulaceae. *Pflanzenreich* 4(237): 17–160.

PAX, F. (1909). Über einen neuen Primulaceen – Typus aus Persien. *Schles. Geesellsch. vaterl. Cul. II. Abt. Naturw. b. Zool.-bot. Sekt. Jahresb.* 87: 19–21.

ROEMER J. J. & SCHULTES, J. A. (1819). *Systema vegetabilium*. 4. Stuttgardiae.

SMITH, W. W. & FORREST, G. (1928). The sections of the genus Primula. *Not. Roy. Bot. Gard. Edinb.* 16: 1–50.

SMITH, W. W. & FLETCHER, H. R. (1948). The genus Primula: Sections Cunneifolia, Floribundae, Parryi and Auricula. *Trans. Roy. Soc. Edinb.* 61: 631–686.

SMOLIANINOVA, L. A. (1952). *Dionysia in Flora SSSR*. 18: 208–217, 728.

SMOLIANINOVA, L. A. & KAMELINA, O. P. (1972). Chromosome numbers of endemic species of *Dionysia* Fenzl (Primulaceae) from Western Gissar (USSR). *Bot. Zurn.* 57: 244–46.

STAPF, O. (1913). Dionysia lamingtonii. *Bull. Misc. Inform., Kew* 1913: 43–44.

WATT, G. (1891). Report of the botanical collections made in south-west Persia by Major H. A. Sawyer. *Report of a reconnaissance in the Bakhtiari country SW Persia*. Appendix A. Simla.

WENDELBO, P. (1958). Primulaceae in Køie, M. & Reichinger, K. H. Symbolae Afghanicae. *IV. Biol. Skr. Dan. Vid. Selsk*. 10(3): 63–76.

WENDELBO, P. (1959). Two new species of Dionysia from Afghanistan. *Bot. Not.* 112: 495–501.

WENDELBO, P. (1961). A Monograph of the Genus Dionysia. . *Årbok* Univ. I. Bergen. Mat.-Nat. Ser. 3: 1–83.

WENDELBO, P. (1964). The Genus Dionysia in Afghanistan with descriptions of 6 new species. *Ibid*. 19: 1–28.

WENDELBO, P. (1965). Ed. Rechinger, K. H. *Primulaceae* for *Flora Iranica* 9/31, 3: 1–37.

WENDELBO, P. (1970). New and noteworthy species of Primulaceae from the 'Flora Iranica' area. *Bot. Not.* 123: 300–309.

WENDELBO, P. (1971). On xeromorphic adaptations in the genus Dionysia (Primulaceae). *Ann. Naturhist. Mus. Wien*. 75: 249–254.

WENDELBO, P. (1976). Another new Dionysia (Primulaceae) from the Bakhtiari Mts of Iran. *Iranian Journ. Bot.* 1(1): 71–73.

WENDELBO, P. (1980). New species of Dionysia, Kickxia and Onobrychis from Iran. *Notes Roy. Bot. Gard. Edinb.* 38(1): 105–110.

Index

Synonyms in *italics*; *d* indicates a drawing; **p** indicates a colour photograph.